Wind and Solar Energy in Agriculture

Wind and Solar Energy in Agriculture

Edited by:

Sujata Nagnath Mustapure

Delve
PUBLISHING

www.delvepublishing.com

Wind and Solar Energy In Agriculture

Sujata Nagnath Mustapure

Delve Publishing
224 Shoreacres Road
Burlington, ON L7L 2H2
Canada
www.delvepublishing.com
Email: orders@arclereducation.com

ISBN: 978-1-77469-126-7 (Hardcover)

ABOUT THE EDITOR

Dr. Sujata Nagnath Mustapure is a teaching associate, Department of Electrical & Other Energy sources, College of agricultural engineering & Technology, Vasantrao Naik Marathwada Krishi Vidhyapeet, Parbhani. She obtained her B.Tech (agril. Engg.) in 2010 from Vasantrao Naik Marathwada Krishi Vidhyapeet, Parbhani. Gold medal in M.Tech (Agril. Engg.), 2014 from Dr. Panjabrao Deshmukh Krishi Vidyapeeth, Akola and Ph.D in Renewable energy engineering in 2019 from Maharana Pratap University of Agriculture and Technology, Udaipur. She has also worked as assistant professor in College of Agricultural, Naigoan, Vasantrao Naik Marathwada Krishi Vidhyapeet, Parbhani. She has authored several national and international publications.

TABLE OF CONTENTS

LIST OF FIGURES

LIST OF TABLES

LIST OF ABBREVIATIONS

AC	Alternating Current
AI	Artificial Intelligence
BCE	Before the Common Era
BTU	British Thermal Unit
CH_4	Methane
CO_2	Carbon Dioxide
CSP	Concentrating Solar Power
CWC	Central Warehousing Corporation
DC	Direct Current
DOE	Department of Energy
EAIS	East Antarctic Ice Sheet
EV	Electric Vehicle
FACE	Free-Air Carbon Dioxide Enrichment
FAO	Food and Agriculture Organization
FCI	Food Corporation of India
FDA	Food and Drug Administration
FHA	Forced Hot Air
GHG	Green-House Gas
GIS	Greenland Ice Sheet
GIZ	Gesellschaft Für Internationale Zusammenarbeit
GM	Genetically Modified
GPS	Global Positioning System
GW	Gigawatt
HFCs	Hydro Fluorocarbon
HP	Horsepower
HVAC	Heating Ventilation and Air Conditioning
IEC	International Electrotechnical Commission

IoT	Internet of Things
IPCC	Intergovernmental Panel for Climate Change
IPM	Integrated Pest Management
IPP	Independent Power Production
IR	Infrared Radiation
ISO	International Organization for Standardization
KW	Kilowatt
LCD	Liquid Crystal Display
LCOE	Levelized Cost of Electricity
LDC	Livestock Development Council
LP	Liquified Petroleum
MW	Megawatt
N_2O	Nitrous Oxide
NREL	National Renewable Energy Laboratory
Op-Amp IC	Operational Amplifier IC
PACEST	Presidential Advisory Council on Education, Science, and Technology
PAEGC	Powering Agriculture: An Energy Grand Challenge for Development
PFCs	Perfluorocarbon
PV	Photovoltaic
PVWPS	Photovoltaic Water Pumping System
R&D	Research and Development
RFID	Radio-Frequency Identification
RTK	Real-Time Kinematic
RuBisCO	Ribulose-1,5-Bisphosphate Carboxylase-Oxygenase
SEIA	Solar Energy Industries Association
SF_6	Sulfur Hexafluoride
SMEs	Small and Medium-Sized Enterprises
SPIS	Solar Powered Irrigation Systems
SPMI	Solar Powered Micro Irrigation
SPV	Solar Photovoltaic
SPVPS	Solar Photovoltaic Pumping System
SPVWPS	solar PVWPS
SRES	Special Report on Emission Scenario
SWC	State Warehousing Corporation

UNEP	United Nations Environment Program
UV	Ultraviolet
VFD	Variable Frequency Drive
WAIS	West Antarctic Ice Sheet
WaPOR	Water Productivity Open-access Portal
WMO	World Meteorological Organization
Wp	Watt-Peak
WUE	Water Use Efficiency

GLOSSARY

A

Abolished – to end a law or system officially.

Accommodate – (of a building or other area) provide lodging or sufficient space for.

Adaptation – the action or process of adapting or being adapted.

Agribusiness – is a term used to describe the sector that encompasses all economic activities that are related to farming, i.e., chemicals, breeding, crop production/farming, farm machinery, distribution, marketing.

AlternatingCurrent – is an electric current that periodically reverses direction, in contrast to direct current which flows only in one direction.

Assimilation – the process of taking in and fully understanding information or ideas.

Auxiliary – providing supplementary or additional help and support.

C

Catastrophic – involving or causing sudden great damage or suffering.

Civilizations – the stage of human social and cultural development and organization that is considered most advanced.

Contaminated – contamination is the presence of a constituent, impurity, or some other undesirable element that spoils, corrupts, infects, makes unfit, or makes inferior a material, physical body, natural environment, workplace, etc.

Convectional – the movement caused within a fluid by the tendency of hotter and therefore less dense material to rise.

Conveyed – transport or carry to a place.

Corporate Social Responsibility – Corporate social responsibility is a type of international private business self-regulation that aims to contribute to societal goals of a philanthropic, activist, or charitable nature by engaging in or supporting volunteering or ethically-oriented practices.

Cultivation – is the act of caring for or raising plants.

D

Desalination – is a process that takes away mineral components from saline water.

Direct Current – is the unidirectional flow of an electric charge. An electrochemical cell is a prime example of DC power.

Disconcerting – causing one to feel unsettled.

Dissemination – the action or fact of spreading something, especially information, widely.

Diversification – the action of diversifying something or the fact of becoming more diverse.
Drip Irrigation – is a type of micro-irrigation system that has the potential to save water and nutrients by allowing water to drip slowly to the roots of plants, either from above the soil surface or buried below the surface.

E
Emission – an emission is something that has been released or emitted into the world.
Environment – The conditions in which you live, work, etc.
Eradicated – (of a tree or plant) depicted with the roots exposed.
Evapotranspiration – the process by which water is transferred from the land to the atmosphere by evaporation from the soil and other surfaces and by transpiration from plants.

F
Food Technology – It is a branch of food science that deals with the production, preservation, quality control, and research and development of the food products.
Furnaces – an enclosed structure in which material can be heated to very high temperatures, e.g., for smelting metals.

G
Gradient – an inclined part of a road or railway; a slope.
Grasslands – are areas where the vegetation is dominated by grasses (Poaceae).
Greenhouse – a greenhouse is a structure with walls and roof made chiefly of transparent material, such as glass, in which plants requiring regulated climatic conditions are grown.
Grid – electrical grid, electric grid, or power grid, is an interconnected network for delivering electricity from producers to consumers.
Groundwater Management – management of groundwater sub-basins to provide for multiple long-term benefits without resulting in or aggravating conditions that cause significant economic, social, or environmental impacts, such as long-term overdraft, land subsidence, ecosystem.

H
Harvesting – The amount of grain, fruit, etc., that is collected.
Horsepower – is a unit of measurement of power, or the rate at which work is done, usually in reference to the output of engines or motors.
Hulls – a hull is the watertight body of a ship or boat.

I
Imperative – of vital importance; crucial.
Innovation – The practical implementation of ideas that result in the introduction of new goods or services or improvement in offering goods or services.
Inscription – a thing inscribed, as on a monument or in a book.

Integrated Pest Management – It is a broad-based approach that integrates practices for economic control of pests.
Intermittently – at irregular intervals; not continuously or steadily.
Inundations – an overwhelming abundance of people or things.
Irrigation – is the process of applying controlled amounts of water to plants at needed intervals.

J
Jammed – squeeze or pack tightly into a specified space.

K
Kinetic – in physics, the kinetic energy of an object is the energy that it possesses due to its motion.

L
Larvae – the active immature form of an insect, especially one that differs greatly from the adult and forms the stage between egg and pupa, e.g., a caterpillar or grub.
Livestock – is commonly defined as domesticated animals raised in an agricultural setting to produce labor and commodities such as meat, eggs, milk, fur, leather, and wool.

M
Monocrystalline – monocrystalline silicon, more often called single-crystal silicon, in short mono c-Si or mono-Si, is the base material for silicon-based discrete components and integrated circuits used in virtually all modern electronic equipment.

N
Net Metering – It is a billing mechanism that credits solar energy system owners for the electricity they add to the grid.

P
Pasture – land covered with grass and other low plants suitable for grazing animals, especially cattle or sheep.
Peak Wattage – this value specifies the output power achieved by a Solar module under full solar radiation (under set Standard Test Conditions).
Photon – is a type of elementary particle. It is the quantum of the electromagnetic field, including electromagnetic radiation such as light and radio waves, and the force carrier for the electromagnetic force.
Photovoltaics – It is the conversion of light into electricity using semiconducting materials that exhibit the photovoltaic effect, a phenomenon studied in physics, photochemistry, and electrochemistry.
Polycrystalline – polycrystalline silicon, or multi-crystalline silicon, also called polysilicon or poly-Si, is a high purity, polycrystalline form of silicon, used as a raw material by the solar photovoltaic and electronics industry.

Precipitation – the action or process of precipitating a substance from a solution.
Proprietorships – the state or right of owning a business or holding property.

R

Rankine Cycle Engine – is a model used to predict the performance of steam turbine systems.
Real-Time Kinematic – It is a technique used to increase the accuracy of GNSS positions using a fixed base station that wirelessly sends out correctional data to a moving receiver.
Renewable – renewable energy is useful energy that is collected from renewable resources, which are naturally replenished on a human timescale, including carbon-neutral sources like sunlight, wind, rain, tides, waves, and geothermal heat.
Rotor – the rotor is a moving component of an electromagnetic system in the electric motor, electric generator, or alternator.
Rudimentary – relating to an immature, undeveloped, or basic form.

S

Semiconductors – a semiconductor material has an electrical conductivity value falling between that of a conductor, such as metallic copper, and an insulator, such as glass.
SolaTube – is a daylighting system that are engineered to efficiently capture the sun's rays and bring that natural light into the house.
Solar Pathfinder – It is the professional PV installer's tool for determining the most economical and efficient photovoltaic array location and position.
Solar Ponds – a solar pond is a pool of saltwater which collects and stores solar thermal energy.
Spatial – relating to, occupying, or having the character of space.
Sprinkler – is a device used to irrigate agricultural crops, lawns, landscapes, golf courses, and other areas.
Substitution – the action of replacing someone or something with another person or thing.

T

Thermostats – is a component that senses the temperature of a physical system and performs actions so that the system's temperature is maintained near a desired setpoint.
Thresher – a threshing machine or a thresher is a piece of farm equipment that threshes grain, that is, it removes the seeds from the stalks and husks.
Trajectory – the path followed by a projectile flying or an object moving under the action of given forces.
Transmission – the action or process of transmitting something, or the state of being transmitted.
Transpiration – (of a plant or leaf) the exhalation of water vapor through the stomata.

V

Variable Frequency Drive – is a type of adjustable-speed drive used in electro-mechanical drive systems to control AC motor speed and torque by varying motor input frequency and voltage.

Veterinary – relating to the diseases, injuries, and treatment of farm and domestic animals.

W

Weevil – a small beetle with an elongated snout, the larvae of which typically develop inside seeds, stems, or other plant parts. Many are pests of crops or stored foodstuffs.

PREFACE

Renewable energy is now powering or assisting several numbers of on-farm necessity of energies, and that varies from water pumping to space heating. Both, farmers as well as ranchers are now more and more selling energy and using these energies into the industry of agriculture (for example, application of solar energy and wind energy).

This is contributing to greater energy security in the field of agriculture with the help of elevated variety of sources of renewable energy, more self-supply of energy to the fields and barns, and also helping in decreasing the environmental effects.

This book takes the readers through an overview of agriculture, its history, and their types. This book sheds light on the significance of renewable energy in the industry of agriculture and various applications of these renewable energy (solar energy and wind energy) in agriculture.

The first chapter stresses on the basic overview of the agriculture, and the history of agriculture so that the readers are clear about the philosophies behind that form the utmost basics in the field. This chapter will also emphasize the importance of agriculture to various aspects, types of various agriculture forms, and also the use of renewable energy in agriculture.

The second chapter takes the readers through the concepts of solar energy, conversion of solar energy to use in the agriculture industry. This chapter will provide highlights on the various key components of a solar-powered system, advantages, and disadvantages of solar energy.

Then, the third chapter explains the potential use of solar energy in the agribusiness. It also explains the technologies based on solar energy such as photovoltaic (PV) energy, and solar water heater systems. This chapter also sheds light on the concept of green technology.

The fourth chapter introduces the readers to the benefits and risks of solar-powered irrigation. This chapter also explains the increasing interest in solar-powered irrigation systems (SPIS), as well as the components of the solar-powered irrigation systems.

The fifth chapter throws light on the significance of wind energy along with the advantages and challenges of wind energy. This chapter explains different types of technologies used to take benefit of wind energy.

The sixth chapter takes the readers through applications of wind energy in the field of agriculture. The readers are then told about the impacts of wind energy on the crops, such as physiological impacts and mechanical impacts, and challenges that have been faced by the use of wind power.

The seventh chapter explains the challenges that have been addressed in the field of agriculture that emphasize lack of availability of good quality seeds, soil erosion, scarcity of capital. This chapter also explains the various innovations that took place in the field of agriculture.

The last chapter of this book sheds light on the impacts of climate change on the agriculture. This chapter also mentions that how climate change will alter the food, how will the global food production be affected in the future by climate change.

This book has been designed to suit the knowledge and pursuit of the researcher and scholars and to empower them with various aspects of agriculture and the role of renewable energies in agriculture so that they are updated with the information. I hope that the readers find the book explanatory and insightful and that this book is referred by the scholars across various fields.

Chapter 1

Introduction to Agriculture

CONTENTS

This chapter sheds light on the basic significance of agriculture. This chapter also explains the basic concept or ideology behind agriculture and different aspects of the history of agriculture such as prehistory, early civilization, early to modern civilization, modern farming, and the ecological age.

This chapter also explains the various types of agriculture and farming that include subsistence farming, commercial farming, nomadic farming, livestock farming, rudimentary sedentary tillage, Mediterranean agriculture, dairy farming, and specialized horticulture.

This chapter also addresses the importance of agriculture concerning the source of livelihood, contribution to the national revenue, supply of food as well as fodder, significance to the international trade, marketable surplus, food security, great employment opportunities, and economic development. This chapter emphasizes the application of renewable energy in agriculture, such as solar energy and wind energy.

1.1. WHAT IS AGRICULTURE?

The term 'agriculture' is derived from a Latin word- 'ager' or 'agri' and culture. The meaning of ager or agri is soil, and the meaning of culture is the cultivation of the soil. In the present term, agriculture means the art and science of cultivating the soil, growing crops, and rearing livestock.

Farming can be considered as a whole system that contains input, processing, and output. The farming inputs are seeds, fertilizer, and equipment. These inputs then go through the processes like plowing, sowing, irrigation, weeding, and harvesting. And after that, the last output comes like crops, dairy products, and poultry products.

In the present scenario, subsistence and commercial are two major verticals of agriculture, these normally belong to the less developed and more developed areas. One of the major ways to differentiate between more developed and less developed areas is the method people by which get the food, which they require to live (Figure 1.1).

Figure 1.1: Introduction to agriculture.

In developing countries, the major population is farmers, and they are producing the food by which they and their families can survive. The scenario is opposite in the case of North America, where less than 5% of the population are a farmer. These 5% farmers can produce sufficient to feed the remaining population of North America and also produce a large amount of additional food.

Subsistence agriculture is commonly established in less developed countries, where the production of food is mainly consumed by the farmer and their family. In subsistence agriculture, small scale farming is mainly for survive the farmer and their family. Rarely there is some extra production of food, it might be sold, but extra production of food is not common. The main objective of commercial farming is to make income.

Intensive subsistence agriculture is the most abundant type of agriculture experienced all over the world, subsistence agriculture is mainly dependent on animal power and usually experienced in the humid and tropical region of the world.

This kind of farming is verified by various struggles to adapt the landscape to the growth production of food. As the word indicates, this type of subsistence agriculture is very much labor-intensive on the farmer using restricted space and restricted waste.

This type of agriculture is generally adapted in East Asia, South Asia, and Southeast Asia, where very high population densities and land use are limited. The wet rice fields are the most common type of this, but could also cover non-wet rice fields like wheat and barley. A method called double cropping, adapted by farmers to efficiently get two harvests per year from a single field, in sunny locations and long growing seasons.

Shifting cultivation is another type of subsistence agriculture because the farmer shifts around to new sites after few years to farm new land. Farming on an area of land tends to reduce its production capacity and the land which is highly fertile after it is first cleared drops its fertility all over several harvests. Shifting cultivation was a common type of farming, in the first agricultural revolution.

There are two methods in shifting cultivation.

- Slash-and-burn agriculture: Farmer must eliminate and burn the earth in a way called slash-and-burn agriculture in which slashing the land clear field while burning the natural vegetation fertilize the soil; and
- Farmer can only produce their crops on the cleared area for 2–3 years, till the soil is reduced of its nutrients then the farmers must transfer and remove a new location of earth, the farmer may come back to the prior land after 5–20 years after the natural vegetation has regrown.

Corn, millet, and sugarcane are the most common crops produced in shifting cultivation. Another traditional quality of LDC (livestock development council) is the subsistence farmers do not own the land, in its place, the chief of village or council controls the earth. Slash-and-burn agriculture has been an important reason to deforestation all over the world. To report deforestation and the safety of species, people required to find out the origin reason such as poverty and hunger.

Pastoral nomadism is the same as subsistence agriculture exclude that the main attention is on domesticated animals rather than crops. Commonly pastoral nomads happen in arid areas such as the Middle East and Northern Africa since the atmosphere is very dry for subsistence agriculture.

The providing of milk, clothing, and tents are the main reasons for raising the animals. The interesting fact with pastoral nomads is that most do not slaughter their herd for meat, most eat grain by dealing milk and clothing for grain with farmers of local area.

The type of animal selection by pastoral nomad mainly depends on the culture of the area, the prestige of animals and the atmosphere. For example, camels can transport a weighty load and travel long distance with very low quantity of water, an important benefit in arid areas. Goats require a high quantity of water but can eat a large variety of food than the camel.

Generally, people think that nomads walk randomly all over the area to get the water, but this is false. In its place, pastoral nomads have great knowledge about their area. The every group of pastoral nomads control a defined area and will hardly occupy another area.

Every area has a tendency to be sufficiently large to surround the required quantity of water and foliage for survival. There are some nomad groups who transfer seasonally between mountainous and low-lying regions, and this process is called transhumance.

The second agricultural revolution and Industrial revolution happen together; this revolution would change agriculture beyond subsistence to produce the type of surpluses required to feed thousands of persons doing jobs in factories in place of agricultural fields.

Farming techniques and machinery improvements that happened in the late 1800s and early 1900s led to healthier diets, longer life expectancy, and facilitate sustain the second agricultural revolution. The development of the railroad helped transfer agriculture into new areas, like United States Great Plains.

Geographer John Hudson marked out the main role railroads, and agriculture played in altering the landscape of that area from open prairie to individual farmsteads. After that, the mechanization of machinery and the development of tractors, combines, and a multitude of large farm machinery, be thinkable by the internal combustible engine. The use of banking services like net banking and lending practices facilitated farmers afford new machinery.

Johann Heinrich von Thunen in the 1800s experienced the second agricultural revolution firsthand. Due to that, he further developed his model named the Von Thünen Model, which is usually described as the first effort in analyzing the spatial character of economic activity. This is considered the birth of commercial agriculture.

Many developed countries tend to have commercial agriculture, having an aim of producing food for sale in the global market place known as agribusiness. In commercial agriculture, the food is rarely sold directly to the consumer. Instead of that, the food is sold to a well-known food-processing company where it is further processed into a complete product. This consists of produce as well as food products.

A key difference between developing countries and developed countries in terms of agriculture is the availability of the work force at that farm. In developing countries, it is common that over half of the work forces are subsistence farmers.

In developed countries such as the US, the workforce that is farmers are much lesser than half. In the US alone, less than 2% of the workforce are farmers, still have skills, knowledge, and advance technology to feed the whole country.

One of the main reasons why just 2% of the US workforce can feed the whole nation has to do with machinery. As machinery use can harvest crops at much large scale and very quickly too. Developing countries have access to transportation networks also, to provide perishable foods such as dairy long distances in a very short period.

Commercial farmers, usually depend on modern scientific improvements so as to produce higher yields, including crop rotation, fertilizers, and herbicides and, hybrid plants and animal breeds. Another kind of commercial agriculture is found in warm and tropical climates. This is known as plantations. A plantation further can be described as a large-scale farm that mainly focuses on the production of single crops like tea, coffee, tobacco, rubber, cotton, and sugar cane.

Such forms of farming are generally found in low developed countries but are usually owned by many corporations in developed countries. In addition, plantations tend to import workers and provide food, water, and shelter provisions for workers to live there for a specific period.

1.2. THE HISTORY OF AGRICULTURE

1.2.1. Prehistory

In ancient times, humans were hunter-gatherers, meaning that people used to live a nomadic lifestyle and move along with the seasons to follow the supply of food. As the glaciers retreated, patterns as well as the growth of plant life changed. It means that the requirement to move became less important. But the lifestyle carried on for so many years as people tried to increase their acquisition of resources.

At present, the term "hunter-gatherer" can be considered as a contradiction because it is believed that nomadic communities did not just decide one day and stop moving and being permanently settled, but the reasons for the transition have been highly explored but still not settled.

Also, it is likely to be possible that they had herds of livestock that they carry along wherever they go or shift, preserving them for further use, such as milk, meat, and various other resources. The need to secure stores of food for coming seasons may have played a major role at that time.

Paleontologists as well as archaeologists have found the origins of farming to approximately 10,000 years back, to somewhere in the Indus Valley and as a distinct development in China besides the Yangtze River.

1.2.2. Early Civilization

Early civilization is regarded as a rich time in the field of agricultural science and technology. Around 5500 BC, the Sumerian civilization of the Middle East and pre-Greco-Roman civilizations understood the need for a specialized workforce in the field of agriculture for their societies to flourish.

It is at that time that people saw the development of irrigation technology along with the development of the specialized workforce. The focus on agricultural technology eventually allowed the growth and development of the first cities.

Agriculture can be regarded as the key reason behind civilization, including social attitudes, legal codes, and religious practices. Due to an agricultural system, Ancient Egypt had the most intricate society of the ancient world before the development of the Greek civilization.

The religious structure and civilization were built around the life-giving nature of the River Nile. The reliance of the kingdom was largely on the river and delta systems. Most of the areas of Egypt were and yet are desert with some retreats which basically means that the process of agriculture was centered on a very small part of the world (Figure 1.2).

Figure 1.2: The history of agriculture.

1.2.3. Early to Modern Civilization

The Middle East countries continued to see a high level of innovation in the industries that were associated with agriculture. The historians refer this to as "The Arab Agricultural Revolution." That was mainly because of the diversity of the local topographies as well as the crops produced in the Middle East and Indus Valley.

A very little changed in Europe prior to the development of the kingdom in the 11th century. The systems usually originated through feudalism, whether church holdings and secular lords, needed to improve the produce along with the gradually increasing population. The effective technological developments have also been observed in that period.

That was a time of huge selective crossbreeding mainly in the animal livestock. In certain parts of Europe, few fragments of the agricultural system can be seen at present in the form of furrow strips and medieval ridge farming.

Wild woodlands were later converted to the plains and floodplains were drained, bracken was cleared for pasture and areas having low fertility were converted to make efficient use of them. For example, barley grows quite well on the salty soil and consequently makes it perfect to grow on the floodplains.

1.2.4. Modern Farming

It was the time around the 18th century when modern farming started. That period is usually referred to as The British Revolution. At that particular time, so many advancements were made in the field of farming, that too in a very short period. By using modern farming methods or techniques, an increase in production was observed.

Later, a four-field system replaced the three-field system of crop rotation. An inclusive enclosure act regulated the management of land and selective crossbreeding started at an industrial level to increase the size of a crop along with its production.

In addition, animal husbandry also improved, which led to a large amount of production as compared to the amount of production under the traditional system. Such kind of modifications allowed the industrial revolution along with the clear concentration of urban development.

At the time when the Corn Laws in England were abolished, it was the mere beginning of the global food economy. Simultaneously, Charles Darwin's Theory of Evolution put the process of agriculture on the modern way of science as people started understanding the growth as well as the development of crops.

Further modifications would not be until the years of war when the military powers developed the idea of intensive farming. Since the year of 1970, the global production of cereals has multiplied.

1.2.5. The Ecological Age

Currently, sustainability is known as the catchword as people try balancing different conflicting requirements, to protect the environment and also to initiate such practices that secure it while providing for the growing requirements of the future population growth.

At present, the food supply feeds around 7 billion people, still there is enough amount of food that can serve even more people. It has been observed that some of the practices have had negative effects on the land. A lot many trees are being cut down for agricultural production, and the soil are becoming more exhausted while distrusting the ecosystems.

Also, people are still agreeing on the universal standards, and they are trying to redefine the factors of what is agriculturally and ecologically sound, providing to the needs of the crop while not impacting the future generations. Due to advancements in agricultural science and technology, the field of agriculture may likely be on the edge of another revolution.

1.3. START OF AGRICULTURE

The growth of agriculture has contributed to the rise of civilization over the centuries. Prior to the time when agriculture became widespread, people used to spend most of their lives in search of food, gathering wild plants and hunting wild animals. Almost 11,500 years ago, people learned how to grow cereals and root crops and then settled down to a life based on farming.

Around 2000 years ago, most of the population present on the earth became dependent on agriculture for their living. People started growing their crops, their food. Also, they started breeding as well as herding wild animals.

Adaptation of wild plants and animals for the use of people is known as domestication. Rice or corn was the first domesticated plant. Chinese farmers were cultivating rice as early as 7500 BCE (before the common era).

On the other hand, the first domesticated animals were dogs, that were used for hunting. Goats and sheep were domesticated next. Pigs and cattle were also domesticated. Most of such animals had been hunted for meat and hides. At present, many of those animals are a source of milk, butter, and cheese. Ultimately, people made use of domesticated animals such as oxe to plow, pull, and transportation.

Agriculture positively enabled people to produce surplus food. They could use this extra food when crops failed or trade it for other goods. Availability of food in the large amount allowed people to perform other tasks that were not related to farming.

Agriculture kept formerly nomadic people near their fields and thus, led to the development of permanent and well-settled villages. Further, they linked through trade. New economies were successful in certain areas that cities grew and civilizations developed. The earliest civilizations depending upon the intensive agriculture arose near the Tigris and Euphrates Rivers in Mesopotamia (now Iraq and Iran) and along the Nile River in Egypt.

For so many years, agricultural development was really slow. The fire was one of the earliest agricultural tools. Native Americans made use of fire for controlling the growth of berry-producing plants, which they knew grew quickly after a wildfire.

Farmers cultivated small plots of land manually. They used axes to clear plants and trees, and digging sticks for breaking up and till the soil. Improved tools of farming such as stone, bone, iron, and bronze were developed over time. Also, new methods of storage were developed.

People started building up stocks of food in jars and clay pits for emergency conditions or to use in times of security. In addition, they also started making clay pots and various other vessels for carrying as well as cooking food (Figure 1.3).

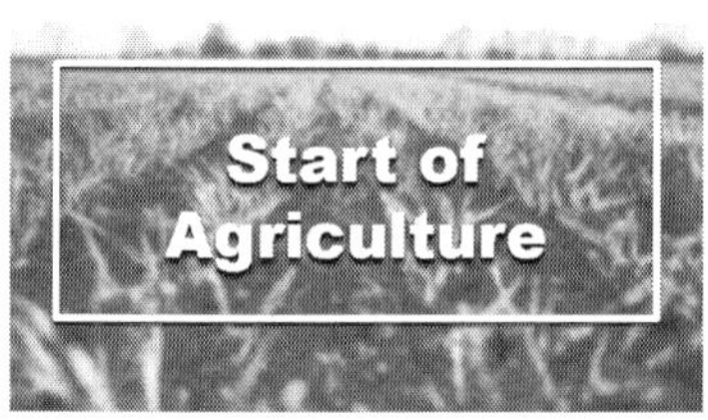

Figure 1.3: Start of agriculture.

Around 5500 BCE, farmers in Mesopotamia developed simple irrigation systems. By channeling water from streams onto their fields, farmers were capable of settling in areas once thought to be unsuited to agriculture. In Mesopotamia, and after that, in China and Egypt, people organized themselves and worked together to construct and maintain better irrigation systems.

Early farmers developed improved varieties of plants. For instance, around 6000 BCE, a new variety of wheat arose in Egypt and South Asia. It was found that the new variety was stronger than the previous cereal grains; its hulls were much easier to remove and also, it could be made into bread.

As the Romans expanded their empire, they improved the best agricultural methods of the people they dominated. They also wrote manuals about the techniques of farming they observed in Asia and Africa, and then adapted them to land in Europe.

Also, the Chinese adopted tools and methods of farming from nearby empires. A large variety of rice from Vietnam quickly ripened and allowed farmers to harvest different crops during a period of a single growing season. This variety of rice quickly became popular all over China.

A lot many medieval European farmers made use of an open-field system of planting. One field used to be planted in spring, another in autumn and the rest left unplanted, or fallow. Such a system preserved nutrients in the soils and as a result of which production of the crop increased.

The leaders of the Islamic Golden Age (which reached its height around 1000) in North Africa and the Middle East made agriculture into a science. Islamic Golden Age farmers learned crop rotation. During the 15th and 16th centuries, explorers introduced novel varieties of plants as well as other agricultural products into Europe.

From Asia, they carried tea, coffee, and indigo (a plant used to make blue dye). From the Americas, they took plants like tomatoes, potatoes, beans, corn, tobacco, and peanuts. Some of these became staples and expanded the diet of people.

1.3.1. Machinery

A period of critical agricultural development started during the early 1700s for Great Britain and the Low Countries (Belgium, Luxembourg, and the Netherlands, which lie below sea level). Novel agricultural inventions intensely increased the production of food in Europe as well as European colonies, mainly the United States and Canada.

One of the key developments in the field of agriculture was an improved horse-drawn seed drill which was invented by Jethro Tull in England. Until that time, farmers used to sow seeds manually, by hand. Tull's drill made rows of holes for the seeds. By the end of the 18th century, seed drilling was largely practiced by Europeans.

A lot many machines were developed in the United States. The cotton gin was invented by Eli Whitney in the year 1794. That machine decreased the time required for separating the cotton fiber from the seed. During the 1830s, Cyrus McCormick's mechanical reaper aided in modernizing the process of grain cutting.

At the same time, John and Hiram Pitts introduced a horse-powered thresher that reduced the time required in the process of separating grain and seed from chaff and straw. John Deere's steel plow, which was introduced in the year 1837, made it easily possible to work the tough prairie soil with very less horsepower (HP).

Along with the development of new machines, there were different key advances in the methods or techniques of farming. Farmers increased the size as well as productivity of their livestock by adopting the method of selective breeding.

Cultures have been breeding animals for centuries. Various evidence suggest that Mongolian nomads were selectively breeding horses in the Bronze Age. In addition, Europeans started practicing selective breeding on a large scale from the beginning of the 18th century. The Leicester sheep is an early example of this, it is an animal that is selectively bred in England for its quality meat as well as long, coarse wool.

Similarly, plants could also be selectively bred for some special qualities. In the year 1866, Gregor Mendel's studies in the field of heredity were published in Austria. He did experiments with pea plants, in which he learned how traits were passed from one generation to the next. His work surfaced the way for the improvement of crops through genetics.

During that time, new crop rotation methods also evolved. Various of these were adopted over the next century or so all around Europe. For instance, the Norfolk four-field system, developed in England, proved to be quite successful.

It involved the annual rotation of various crops, such as turnips, wheat, clover, barley, and ryegrass. These added nutrients to the soil enable farmers to grow a sufficient amount of food to sell some of their products without having to leave any land unplanted.

Most of the people around the world were not affected by such developments, but farmers in Australia, Asia, South America, and Africa continued using old ways or methods of agriculture.

1.3.2. Agricultural Science

During the early 1900s, an average farmer in the U.S. produced surplus food to feed their families. On the other hand, a lot of today's farmers can easily feed their families along with thousands of people. The question arises, how did this great leap in productivity come about? It happened mainly because of scientific advancements as well as the development of new power sources.

In developed countries, most of the farmers by the late 1950s were using both gasoline as well as electricity to power machinery. Draft animals and steam-powered machinery were replaced by tractors. Farmers were using machines at almost every stage of cultivation as well as livestock management.

First of all, electricity became the power source of farms in Germany and Japan during the early 1900s. By the year 1960, most of the farms in the US and various other developed nations were electrified. Electricity lit farm buildings and powered machinery such as milking machines, water pumps, and feeding equipment. At present, electricity controls the whole environment in poultry houses and livestock farms.

By tradition, farmers have used a large variety of methods or techniques to protect their crops from different pests and diseases. They have put herb-based poisons on crops, bred strong varieties of crops, handpicked insects off plants and rotated crops to control insects.

These days, almost all farmers, mainly in developed nations, depend on chemicals for pest control. The definition of 'pest' varies from insects to animals like mice and rabbits, and weeds, disease-causing organisms such as fungi, viruses, and bacteria. By using chemicals, losses of crops and prices have dramatically declined.

For so many years, farmers used to depend on natural fertilizers, materials such as wood ash, manure, fish or fish parts, ground bones, and bat and bird waste is known as guano-to increase the nutrients or replenish the soil.

At the start of the 1800s, scientists discovered elements that were most critical for the growth of plants, those were nitrogen, potassium, and phosphorus. After that, fertilizers having these elements were manufactured in the US and Europe. At current time, a lot many farmers use chemical fertilizers with phosphates and nitrates as they increase the crop yields to a greater extent.

Although, fertilizers as well as pesticides have come with another set of key issues. The dependence on chemicals has heavily disturbed the environment, usually by damaging helpful species of animals. In addition, chemical use in large amounts may pose a severe health hazards to people also, mainly via contaminated water supplies.

That is why agricultural scientists are looking for safe or non-hazardous chemicals that can be used as fertilizers and pesticides. Also, some of the farmers make use of natural controls and depend very little on chemicals.

1.3.3. Farming in Water

Agriculture comprises different forms of cultivations-aquaculture and hydroponics. Both of these forms further involve farming in water. Aquaculture is the cultivation of fish and shellfish. It was initially practiced in India, China, and Egypt about thousands of years ago. Now, it is used in ponds, lakes, oceans, and some other water bodies all around the world. Some forms of aquaculture, for example, shrimp farming, have become critical industry in many Asian and Latin American countries.

On the other hand, hydroponics is known as the science of growing plants in the nutrient solutions. Only one acre of nutrient solution can produce more than 50 times the amount of lettuce that is grown on the same amount of soil.

Climate change as well as advancement in technologies are changing the way by which freshwater and ocean fisheries operate. Global warming has largely pushed warm-water species towards the poles and declined the cold-water habitat species. Traditional communities of fishing in both developed as well as developing countries find the number of fish declines.

Bottom trawling has severely affected the ecosystems present under the ocean. In bottom trawling, enormous nets are quite strong from fishing boats and dragged at the bottom of the ocean. The nets catch halibut and squid, but also stir up the sediment at the bottom of the ocean. This disturbs the marine life that is, plankton, and algae which forms the basis of the food chain.

1.4. TYPES OF AGRICULTURE

Figure 1.4: Types of agriculture.

1.4.1. Subsistence Farming

Farming is done for the consumption of the farm owners. This can either be primitive or intensive. The key aim of this farming is to fulfill the requirements of the farmer as well as his family. The main characteristic of this kind of farming is that nothing is sold off the farm.

It has been very common in areas of middle latitudes having soil of low fertility, or other areas with rough terrain. However, it has significantly declined after the collectivization, mainly in Russia, which was one of the main regions where this farming was majorly practiced.

Primitive Subsistence Farming: It is a kind of subsistence farming that is usually done on small areas of land. It uses traditional tools such as Dao, hoe, digging sticks, etc. Rather, it is the most natural method of producing crops, the reason being the natural environment such as rain, heat, wind, and condition of the soil contribute to the growth of crops.

Further, primitive farming includes shifting cultivation. In shifting cultivation method, farmers make the cultivated land clear after harvesting the crops by burning the land. Consequently, they maintain the soil fertility, so whomsoever utilizes the land next can produce a good yield. This method is known by various names in different regions. Shifting cultivation is practiced in some countries of Southeast Asia and South America.

In general, this kind of farming is adopted in the tropics. Under this process, the land is made clear by adopting the method of deforestation (clearing the forests) by using a slash and burn technique. After that, the land is cultivated for some years, or until the fertility decreases or the land is overtaken by the weeds and some other native flora. At this point, farmers move on to clear another area of the forest.

This subsistence kind of farming is almost done manually, most of the time. It is generally adopted by people living in tropical areas such as Southeast Asia, with a key emphasis on grain crops. However, this type of subsistence farming is gradually declining because of the pressure by environmentalists.

1.4.2. Commercial Farming

Commercial farming is a type of agriculture in which farmers grow crops and rear animals for economic activity. Farmers make large use of machinery and cultivate larger areas of the land because of the need for a high amount of output.

However, this is practiced over a small area. This kind of farming is significant in terms of its commercial value. Tropical crops such as tea, coffee, rubber, palm oil, etc. Are some of the major products of this kind of farming.

This kind of farming has developed in different parts of Africa, Asia, and Latin America, where the colonial influence of Europeans has remained. Most of the plantations were developed to provide tropical crops to the European markets. This farming is regarded as a highly capital-intensive type of farming and most of the crops are tree crops.

Commercial farming has three main categories:

- **Commercial Grain Farming:** In this method, as the name suggests, farmers grow grains and trade them in the market. Maize and wheat are the most common crops of commercial grain farming. Usually, farmers of Europe, Asia, temperate grasslands of North America practice this kind of farming.

It is a kind of response to farm mechanization, and it is the main type of farming in such areas where the population is less and rainfall too. These crops are inclined to the vagaries of weather and droughts. The monoculture of wheat is the general practice. Prairies, steppes, and temperate grasslands of Australia and South America are the key areas for this type of farming.

- **Plantation Farming:** It is a combination of industry and agriculture. This is practiced across a wide area of land. Plantation owners in general, grow a single crop such as tea, coffee, banana, etc.

They use technological support for processing the crop on the farm itself or in a factory attached to it. The end product works as a raw material for several industries. For instance, the rubber industry makes use of the rubber produced from its plantation as a raw material.

- **Mixed Farming:** This method of farming comprises the cultivation of crops, nurturing their livestock, and growing their fodder. It is a very common practice in some parts of Australia, the USA, New Zealand, Europe, and South Africa to do mixed farming for a living.

1.4.3. Nomadic Herding

It is completely based upon the rearing of animals on natural pastures. This practice is usually performed by people living in semi-arid and arid regions. These people always stay on the move along with their animals in search of natural pastures, especially for their livestock to graze.

In this process, the kind of animals varies from one region to the other. The typical regions where this type of farming is mainly practiced are Northern Africa, parts of Arabia and parts of northern Eurasia. In addition, this is a subsistence type of activity.

1.4.4. Livestock Ranching

Under this kind of farming, the key focus is laid on rearing animals. Different from nomadic herding, here the farmers live a well-settled life. It has developed on a commercial basis in certain parts of the world where a large area of land is easily available for animal grazing. Some of the main areas where this is highly practiced are Australia, North America, and South America. Animals are reared for wool and meat. They are kept on large-scale farms. These farms are often called ranches.

1.4.5. Rudimentary Sedentary Tillage

This is a subsistence kind of agriculture. It is different from the above-mentioned types as the same plot of land is continuously cultivated year after year, in this type of farming. Fallowing of land is generally adopted to maintain the fertility of the soil. This technique is further adopted in tropical regions. Apart from grain crops, some of the tree crops like the Pará rubber tree are also grown using this system.

1.4.6. Intensive Subsistence Farming with Rice Dominant

Intensive subsistence farming is usually done in tropical areas that have high populations and receive a very large amount of precipitation. In this type of farming, rice is considered to be a dominant crop, as it can employ as well as feed a very large number of people per unit area.

This farming is largely practiced in the Southeast Asian region. It requires the use of both manuals as well as animal power. Farmers generally try to improve the productivity per unit of area with the use of manures.

1.4.7. Intensive Subsistence Farming Without Rice Dominant

This is a modified version of the above-mentioned type of farming. Basically, this is designed for such areas where the amount of rainfall is not very high. Such areas grow crops other than rice, like millets and wheat. Besides the reasonably fewer wet areas of Asia, northern Africa and the parts of the Middle East use this type of farming. Also, it is commonly practiced in certain parts of southern Africa and Central America.

1.4.8. Mediterranean Agriculture

The typically rough terrain of the Mediterranean area has eventually resulted in typical combinations of crops and livestock. Vineyards, wheat, and citrus fruits are the main crops, while small animals are the main livestock reared in the region. In such regions, horticulture is the major activity. Most of the crops are grown during winters with the help of winter rains.

1.4.9. Livestock and Grain Farming

In general, it is also known as mixed farming. Basically, it is originated in the humid regions of the middle latitudes except Asia. The development of such farming is closely associated with the market facilities. Well, it is also said that this farming is a European type of farming. This kind of farming is mostly practiced in the areas of New Zealand and Great Britain.

1.4.10. Dairy Farming

This farming tool is originated in Europe, and from Europe, it spread to other areas or regions. A temperate climate and nearby availability of the market are the two major factors that have been highly accountable for the development of such kind of farming. Countries such as Sweden and Denmark have seen the maximum development of this farming.

1.4.11. Specialized Horticulture

This kind of farming has developed to take better advantage of a huge demand for horticulture-related products, mainly in the areas of large-scale urbanization as well as high-density population. It has been very successful when utilized for vineyard cultivation in certain areas of France, northern Hungary, and the Swiss Lakes regions.

However, Whittlesey's agricultural classification is a bit elaborate, the regionalization on the basis of this classification is not permanent. Because of the changing market demands and evolving agricultural technology, a high number of changes are taking place in the agricultural pattern of the whole world since Whittlesey's study.

High demands for fruits as well as vegetables in the urban areas have resulted in modified land use in various parts of the world. Such kinds of factors usually impart a dynamic character to agricultural activity.

1.5. USES FOR AGRICULTURE

Native non-crop plants can provide habitat for natural enemies as well as pollinators. In addition, they also protect the health of waterways (Figure 1.5).

Figure 1.5: Uses for agriculture.

1.5.1. Pollinator Strips

More than 70% of the total crops are bee-pollinated, with a third of food being dependent on pollination. This dynamic provisioning ecosystem service can be continued by planting habitats such as native shrubs and flowering plants. They will provide pollen and nectar all around the growing season.

1.5.2. Natural Enemy Habitats

Insect predators as well as parasites are known as natural enemies. They can effectively control the pest population in agricultural crops. There are chances that the natural enemies grow in undistributed areas that provide overwintering habitat, flowers to support their reproduction as well as survival, and refuge from pesticide applications in different crops.

The population of natural enemies can control the pest populations in nearby crop fields and various other plantings. Natural enemies may be efficiently conserved with the same plantings that support the pollinators.

1.5.3. Conservation Areas and Riparian Buffers

There are certain programs launched by the National Resources Conservation Service along with other organizations that support non-crop plantings conserving soil and reducing pollution and provide wildlife habitat in farming landscapes.

Such areas are often known as conservation plantings if they are in previously farmed croplands. For instance, the areas that are planted to native species along waterways can effectively provide significant ecosystem services of supporting water filtration as well as preventing flooding, and also, for providing habitat for wildlife.

1.5.4. Bioenergy Crops

Fast-growing native perennials have a lot many benefits as bioenergy crops. Many store Carbon below the ground in their roots, aiding in the reduction of greenhouse gasses and can effectively support populations of beneficial natural enemies as well as pollinators. The biomass present above the ground can be harvested and burned to produce cleaner electricity or that can also be processed into biofuels to partially replace gasoline.

1.6. THE IMPORTANCE OF AGRICULTURE

Agriculture has been linked with the production of significant food crops for decades. Currently, agriculture, not just being a practice of farming, includes dairy, forestry, poultry, fruit cultivation, arbitrary, beekeeping, etc.

At present, the processing, marketing, and distribution of food crops as well as livestock products are recognized as a key part of modern agricultural practices. Therefore, agriculture could be defined as the process of production, processing, promotion, and distribution of agricultural products (Figure 1.6).

Figure 1.6: The importance of agriculture.

These days, agriculture is playing a major role in the life of a particular economy. It is considered the backbone of the economic system of every nation. Apart from providing raw material and food, agriculture also provides several employment opportunities to people. Some of the crucial importance of agriculture are briefly described in subsections.

1.6.1. Source of Livelihood

Agriculture has always been considered the main source of livelihood for a large percentage of the population. Approx. 70% of the population directly depends on agriculture as a source of living. Although, the large percentage of people being dependent on agriculture or agricultural practices is a consequence of the non-development of non-agricultural activities that could easily absorb the growing population. Though, a lot many people in developed nations do not get involved in agricultural practices.

1.6.2. Contribution to National Revenue

For a lot many developing nations, agriculture is the main source of national income. On the other hand, for developed nations, agriculture only contributes a very small percentage to their national income. They have various other activities contributing to national income.

1.6.3. Supply of Food as Well as Fodder

It is largely known that; it is the agricultural sector that provides fodder for domestic animals. The cow provides milk to people. In addition to that, livestock also meets the food requirements of many people.

1.6.4. Significance to the International Trade

Agricultural products such as tea, sugar, spices, rice, coffee, tobacco, and many more contribute to the major products of exports of many countries that depend on agriculture. Imports are considerably reduced if there is smooth development of agriculture practice. Eventually, the export will increase.

This will aid the countries to reduce the unfavorable balance of payments and saving foreign exchange. Further, that amount may be used to import some other important inputs, raw materials, machinery, and various other infrastructure that is important for supporting the economic development of any nation.

1.6.5. Marketable Surplus

Marketable surplus is the result or the outcome of the growth of the agricultural sector. As the countries are developing, a lot of many people are getting involved in manufacturing, mining, and various other non-agricultural sectors. All such people depend on the production of food that they might meet from the marketable surplus of the country. As the development of the agricultural sector is taking place, production is also gradually increasing, leading to the increase of marketable surplus, which may eventually be exported to other countries.

1.6.6. Source of Raw Material

Agriculture is the key source of raw materials to major industries like sugar, cotton, and jute fabric, tobacco, edible, and non-edible oil, etc. In addition to that, various other industries like processing of fruits as well as vegetables and rice husking get their raw material primarily from agriculture.

1.6.7. Significance in Transport

Railways and roadways are efficiently used to transport bulks of agricultural products from farms to various industries. Internal trade is in agricultural products, mostly. Furthermore, to a greater extent, the revenue of the government depends on the success rate of the agricultural sector.

1.6.8. Foreign Exchange Resources

The export trade of the nation relies largely on the agricultural sector. For an instance, agricultural commodities like tobacco, spices, jute, raw cotton, oilseeds, coffee, tea, etc., account for around 18% of the total value of exports of any country. This eventually shows that agricultural products continue to be a substantial source of earning a country foreign exchange.

1.6.9. Great Employment Opportunities

Construction and maintenance of drainage system, irrigation schemes and various other related activities play a key role in the agricultural sector as it offers a large employment opportunity. The agricultural sector offers more and more employment opportunities to the labor force. This ultimately decreases the high rate of unemployment in developing nations caused by the fast-growing population.

1.6.10. Economic Development

As agricultural sector employs a lot many people, it also contributes to the economic development of a country. Consequently, the level of national income along with the living standards of people of that nation is improved.

The fast rate of development in the agricultural sector provides an advanced outlook along with high motivation for development. Therefore, it helps in creating the perfect atmosphere for the overall economic development of any nation. Thus, it can be said that economic development depends on the growth rate of the agricultural sector.

1.6.11. Source of Saving

Certain advanced development in the field of agriculture may also result in savings. Rich farmers, today have started saving, especially after the green revolution. This huge quantity may be invested further in the agriculture sector so as to develop the sector.

1.6.12. Food Security

Food security of a country is guaranteed by a stable agricultural sector. Food security is the key requirement of any country. People are well aware that food security prevents malnourishment, which has been considered to be one of the main issues faced by the developing nations. Most of the countries, these days, depend on agricultural products and associated industries for their key source of income.

1.7. RENEWABLE ENERGY IN AGRICULTURE

There is a great potential for agricultural involvement in the production as well as consumption of wind, solar, biomass, and geothermal energy. Renewable resources are abundant. They are distributed widely all around the world. A high number of commercial technologies are available to connect to such resources, and with the availability of apt support, advanced technologies-some potentially paradigm-shifting-could be brought to the market.

There are numerous ways in which it is considered to be the "back-to-the-future" scenario. It includes a movement towards a more self-sufficient farm as well as a key role for agriculture in energy supply. For example, increased renewable energy in as well as from agriculture calls to mind Henry Ford predicting automobiles fueled by alcohol, while windmills powering water pumps.

Renewable technologies, these days are supplying or supplementing various kinds of on-farm energy requirements, from space heating to water pumping. Progressively, ranchers and farmers are selling energy, for example, electricity generated from wind turbines, biofuels, and other products from biomass.

In addition, this is also contributing to greater energy security in the field of agriculture by increased diversity of energy sources, more, and more self-supply of energy and the reduced environmental impact.

Many countries are facing a choice of energy futures. Continuing the present course is one alternative. Fossil energy for mechanized agriculture has been a major driver of the Green Revolution of increasing the productivity of farms (Figure 1.7).

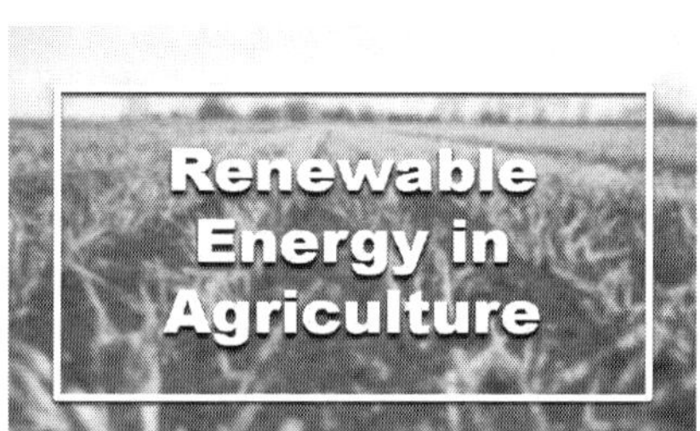

Figure 1.7: Renewable energy in agriculture.

At present, three energy inputs that is, fertilizer, diesel fuel, and electricity is responsible for more than three-quarters of the total farm energy use

(Miranowski, 2004). At projected levels of the production and consumption of oil, America, for example, will be relying on foreign oil imports in the near future, making the country much more vulnerable to disruptions of oil and spikes in prices.

In the field of agriculture, an energy supply disruption of a short time period could mean a significant reduction or even the complete loss of an entire growing season. Farmers, as price-takers for their commodities, are usually not able to pass price increases for fertilizer or energy on to the consumer, and as a result, receive a much lower return for their products when prices increase (Costantini and Bracceva, 2004).

Renewable energy can deal with various concerns that are associated with the use of fossil energy. It produces almost little or no environmental emissions, and also, it does not depend on the imported fuels. Renewable resources are infinite and many of them are available all around the globe.

However, price competitiveness has been a major concern. Costs have significantly decreased since the initial wave of interest in renewable energy in the 1970s. Such technologies at present, provide 6.1 quadrillion British thermal units (Btu) for domestic energy consumption in the US.

In their development, different renewable technologies are at different points. Some of them are commercially available while, on the other hand, others have much potential for a longer period. Unluckily, various advantages that renewable energy can provide are not monetized-which means they cannot be acquired through price signals. Policies are required to pull or push these novel and advanced technologies to full commercial development.

1.7.1. Solar

Solar technologies produce thermal or electrical energy. Photovoltaic (PV) cells (or "solar cells") are the ones that convert sunlight directly into electricity. They are made of semiconductors such as crystalline silicon or various other thin-film materials.

Solar thermal technologies acquire heat from the sun and then further use it directly for water and space heating. Also, convert the heat into electricity via conventional steam cycles, heat engines and some other concentrating solar systems. In the coming years, solar energy could produce hydrogen easily to provide transportation chemicals, fuels, and electricity, and also, to aid as energy storage for the times when the sun does not shine in the sky.

Due to technological advances, the expenses of these technologies have been consistently declining, and high-power expenses can overcome any barrier further. Well, solar resources are the most prominent in the Southwest of the US (around 25% higher than the public normal), sun-based power that is solar electricity, might be more financially effective in states with high power costs.

For example, in New York, the electricity prices can be 50% higher than that of Arizona (U.S. Department of Energy (DOE), Solar Energy Technologies Program, 2003). In the field of agriculture, PV can economically provide electricity where the distance is too high to justify new power lines. Solar electric systems are further utilized for providing electricity for lighting, small motors, battery charging, water pumping, electric fences, etc.

Animals and dairy tasks frequently have considerable water and air heating requirements. For instance, commercial dairy farms utilize a lot of energy to warm water for cleaning hardware or equipment used. Heating water as well as cooling milk can make up to 40% of the energy utilized on a dairy farm.

Solar water heating systems might be utilized for supplying all or just a part of these hot water necessities. Other solar applications incorporate solar crop drying and greenhouse heating (National Renewable Energy Laboratory, n.d.).

The number of solar or sunlight-based energy applications is likely to develop as new advancements in technologies increase the efficiency of solar cells and decreased costs. New "quantum dot" materials could hypothetically more than twofold effectiveness, changing over 65% of the sun's energy into power/electricity, when compared with the best financially accessible solar cells, which have transformation efficiencies of up to 30%. Various research is also being led to decrease the expense of solar water heating systems by using materials such as plastics rather than glass and metals.

1.7.2. Wind Energy

Wind technologies provide mechanical energy as well as electrical energy. The operation of wind turbine is based on a simple principle: rotor blades are turned by the wind, which drive an electric generator, converting the kinetic energy of wind into electric energy.

Basically, the wind is a renewable energy source, and windmills do not generate any injurious environmental emissions. Utility-scale turbines range in size from 750 Kilowatt (KW) to 5 megawatts (MW), out of which major turbine exceeding 1 megawatt (MW). In general, turbines are categorized into wind form, which provides bulk power to the electricity grid.

Small wind turbines usually range in size from 0.4 to 1.5 kW generators for small loads, for example, battery charging for small cabins and sailboats, to 3 to 15 kW systems for a home, to those that create up to 100 kW of power for larger loads, such as small commercial operations.

The technology of wind power is as of now in boundless use because of generous advancement in reducing the costs for regions with reliably high wind speeds. At the end of the year 2005, the wind was accountable for 9,149 MW of electrical producing capacity in the United States.

At an average capacity factor of 31%, this is similar to producing the total amount of electricity that is utilized by more than 2 million American households. There are various commercial wind energy installations all around the world.

Operating in high-wind areas, today's state-of-the-art wind turbines can easily produce electricity for a few cents per kilowatt-hour (kWh), which is competitive with the cost of fossil fuel-fired plants.

Small wind systems can serve farming generally, for example, utilizing mechanical energy to pump water or granulate grain. As costs reduce, small systems used to produce power or electricity may also turn out to be financially proficient by dodging the cost of installing transmission wires, particularly in more far off applications.

Small windmills, when connected to the electricity distribution grid, can effectively generate revenue by sales of electricity when generation surpasses the internal requirements. On the other hand, decentralized wind systems can be connected with other energy sources to develop a hybrid energy system.

In that system, the low cost as well as intermittent wind resource is supplemented by more costly but small generators, for example, batteries or diesel generators, so as to provide power that is reliable as well as inexpensive (Bergey, 2000).

The small wind turbine industry evaluates that 60% of the United States has sufficient wind resources for small turbine use, and 24% of the populace lives in rural areas where drafting and development codes grant installation (National Renewable Energy Technology, 2004).

As technological improvements keep on expanding the economic effectiveness of wind energy, agricultural producers are probably going to increase their utilization of wind power to bring down energy costs and become more energy independent.

1.8. CONCLUSION

In the conclusion of this chapter, the basic concept of agriculture along with its histories such as prehistory, early civilization, early to modern civilization, modern farming, and ecological age has been discussed. This chapter also discussed the start of agriculture, for example, machinery used in agriculture, the science behind agriculture, and water used in farming.

This chapter also discussed the types of agriculture and farming such as subsistence farming, commercial farming, nomadic farming, livestock ranching, rudimentary sedentary tillage, Mediterranean agriculture, livestock, and grain farming, dairy farming, and specialized horticulture.

In this chapter, various important factors of agriculture have been discussed as well, which includes the source of livelihood, contribution to the national revenue, the significance of agriculture with respect to the international trade, the significance of farming to transport, economic development, food security, and source of raw material. This chapter also discussed the role of renewable energy in the field of agriculture, such as the application of solar energy and wind energy in agriculture.

REFERENCES

1. Aqleem Abbas,*Introduction to Agriculture*(2011). (1st edn.) [eBook] Available at: https://www.researchgate.net/profile/Aqleem-Abbas-2/publication/282905517_Introduction_to_Agriculture/links/57e0088b08aee5d2c6dd9eae/Introduction-to-Agriculture.pdf (accessed on 5 May 2021).
2. Choicesmagazine.org. (2006). *Choices Article-Renewable Energy in Agriculture: Back to the Future?* [Online] Available at: https://www.choicesmagazine.org/2006-1/biofuels/2006-1-05.htm (accessed on 5 May 2021).
3. Dastrup, R., (n.d.). *6.2 Types of Agriculture.* [Online] Humangeography.pressbooks.com. Available at: https://humangeography.pressbooks.com/chapter/6-2/ (accessed on 5 May 2021).
4. Farmingportal.co.za. (2019). *The Importance of Agriculture.* [Online] Available at: https://www.farmingportal.co.za/index.php/farminglifestyle/agri-tourism/195-the-importance-of-agriculture (accessed on 5 May 2021).
5. https://www.environmentalscience.org/. (n.d.). *History of Agriculture.* [Online] Available at: https://www.environmentalscience.org/history-agriculture (accessed on 5 May 2021).
6. https://www.toppr.com/. (2021). *Introduction to Agriculture.* [Online] Available at: https://www.toppr.com/guides/geography/agriculture/introduction-to-agriculture/ (accessed on 5 May 2021).
7. National Geographic Society, (2011). *Agriculture.* [Online] Available at: https://www.nationalgeographic.org/encyclopedia/agriculture/ (accessed on 5 May 2021).
8. Native Plants and Ecosystem Services, (n.d.). *Uses for Agriculture.* [Online] Available at: https://www.canr.msu.edu/nativeplants/getting_started/agriculture (accessed on 5 May 2021).
9. Owl Cation-Education, (n.d.). *Types of Agriculture.* [Online] Available at: https://owlcation.com/agriculture/Types-of-Agriculture (accessed on 5 May 2021).

Chapter 2

Solar Energy: A Fundamental Review

CONTENTS

The chapter "solar energy: a fundamental review" explains the basic significance of the production of electric energy with the help of solar radiation. This chapter also explains how the conversion of solar radiation into the form of electricity or electric current takes place.

This chapter provides highlights on the several numbers of various uses of solar energy such as thermal energy, electricity generation, solar water heating, solar ventilation, solar lighting, portable solar, solar transportation, wearable solar tech, along with the other applications.

This chapter also explains the several components of the solar power systems such as the roof system, solar panels, inverter, and net meter. This chapter also provides highlights on the thing that needs to be considered before the installation of the solar power systems, along with this, the chapter also mentions the impact of solar energy on the environment.

This chapter also addresses the importance of solar energy, such as the promise of solar energy, the growth of solar energy, the basics of solar energy. This chapter also includes the advantages and disadvantages of solar energy.

2.1. INTRODUCTION

The definition of solar energy is the radiation that is coming from the Sun, and having the potential of producing heat, triggering chemical reactions, or producing electric energy. The total amount of solar energy that structs the earth is massively in excess of the present as well as the expected necessity of the energy, all across the globe. It suitably harnessed, the highly diffused source has the potential to fulfill the requirement of the energy in the coming interval of time.

Talking about the 21st century, solar energy is anticipated to become more and more attractive in the form of the source of renewable energy. And the reason behind this is its never-ending supply and its non-polluting property, in stark contrast to the finite fossil fuels coal, natural gas, and petroleum (Figure 2.1).

Figure 2.1: An image of solar panel.

Source: Image by Pxfuel.

The Sun is a highly powerful source of the energy, and the solar radiations are by far the largest source of the energy that is received by the earth, but in addition to this, the intensity of the solar radiation or the sunlight at the surface of the earth is quite low. This is mainly due to the enormous radial spreading of the radiation from the huge distance of the Sun.

A comparatively minor additional loss is because of the atmosphere and the clouds of the earth. The atmosphere and the clouds absorb or scatter as much as 54% of the incoming light from the Sun. The solar radiations or light from the Sun reaches the ground is consisting of near about 50% of visible light, near about 45% of IR (infrared radiations), and little amount of the UV (ultraviolet) rays, along with the other kids of electromagnetic radiations.

The Sun does more than for the planet earth despite just provide light through the course of the daytime. In which,every particle of the solar radiation or the sunlight (that is known as a photon) reaches the earth, and is consisting of energy that fuels the earth.

Solar energy is the ultimate source, that is accountable for all of the weather systems and energy sources that are found on the earth, and a sufficient amount of sunlight reaches the surface of the earth each hour to theoretically satisfy the global energy requirement for near about a whole year.

Where does all of this energy come from? The Sun, which is just like any kind of star that is present in the galaxy. The Sun is like a huge nuclear reactor. Deep in the core of the Sun, due to the nuclear fusion reactions, the huge amount of energy generates that radiates in the upward direction from the surface of the earth as well as into space in the form of light and heat.

Solar power can be harnessed and can be transformed in the form of usable energy with the help of PV (photovoltaics) or solar thermal collectors. Nevertheless, solar energy is only responsible for a small amount of overall consumption of energy, the falling cost of installing solar panels means that more and more numbers of people in several places can have the benefits of solar energy. Solar is a resource of renewable energy, it is clean, and in the end, it plays a significant role in the global energy future.

2.2. WHAT IS SOLAR ENERGY?

As per the scientific observations and researches, the earth stops the huge amount of solar energy or solar power, to be specific, it is 173 trillion terawatts. That is 10000 times more power when it is compared with the amount of energy consumption by people. This proves the fact that the Sun is the most abundant source of energy on the entire globe and that it could one day be the most needful source of energy.

Conventionally, all across the globe, the electrical requirements have been satisfied with the help of the fossil fuels such as coal, oil, natural gas, and so on. Nevertheless, these sorts of energy sources have two negative effects:

They play a significant part in global warming as well as acid rain pollution. As an outcome, it negatively affects the several numbers of various animals, plants, along with humans in the environment.

Several nations have full access to the energy resources that are based on fossil fuels, which can take into the direction towards global political as well as economic stability (Figure 2.2).

Figure 2.2: An illustration of solar panels installed on the rooftop.

Source: Image by Pxfuel.

The best substitute is solar energy, which is a renewable resource. In simple words, solar energy will not become unavailable at any time. Solar energy offers a never-ending, steady supply across time. Solar energy is also a green source of energy and the reason behind this that solar energy does not emit pollutants through the course of the process of energy production.

Solar energy is the energy that is generated with the help of the Sun in the form of heat and light. Solar energy is one of the most renewable and readily available sources of energy that exists on the earth. In the matter of fact, solar energy is available in huge amounts and it is free, as well as solar energy does not belong to any person, which makes solar energy one of the most significant non-conventional sources of energy.

Solar energy has been used with the help of people, from ancient times simply with the help of magnifying glass to concentrate the sunlight in the form of beams, in such a way that they would cause wood to catch fire.

Necessarily, solar energy can be utilized to transform it in the form of heat energy, or the solar energy can be transformed in the form of electric current. Solar energy is the energy that is harnessed from the Sun.

2.3. CONVERSION OF SOLAR ENERGY

There are several numbers of various methods to utilize the energy or the solar energy that is coming from the Sun. Out of which, the two main methods to utilize the energy from the Sun are PVs and second is solar thermal capture.

PV is comparatively more common for smaller-scale electricity projects (such as installation of residential solar panels). On the other hand, solar thermal capture is generally only used to the production of electric current on huge scales in utility solar installations.

Moreover, to the production of the electric current, lower temperature fluctuations of the solar thermal projects can be used for heating and cooling as well. Solar energy is one of the rapidly growing as well as cost-effective sources of the power in the world, and it will continue to spread with much more rapid pace in the coming interval of time. With each passing year, the improvements in the solar panel technology, the economic benefits of the solar improve, adding to the environmental perks of choosing a clean, renewable energy source.

2.3.1. How Solar Energy Is Converted to Electricity?

The first and foremost step, to transform the solar energy in the form of electric current, is the installation of PV cells (or photovoltaic cells) or solar cells. The term "photovoltaic" means light and electricity. These solar cells capture the energy coming from the Sun and then transform it in the form of an electric current (Figure 2.3).

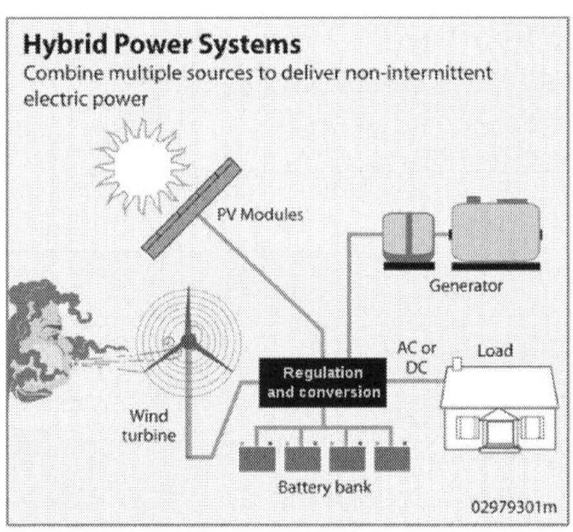

Figure 2.3: An illustration of conversion of solar energy: A hybrid power system.

Source: Image by Wikipedia.

The solar cells or PV cells are made up with the help of material that depicts the PV effect. In simple words, when the solar radiations reach the PV cells, the photons of the light activate the electrons that are present in the solar cells, triggering them to start flowing, and in the end, the production of electric current takes place.

When looking to purchase solar panels, it would be very helpful to know the options that are available in the market. Here is an outline of the main ones, as mentioned below:

- **Polycrystalline:** This utilizes multi-crystalline silicon.
- **Monocrystalline:** This is ideal for small spaces.
- **Thin Film:** Generally, bigger in size and a lot more effective and successful through the course of the day.

The difference between the material that is being used to manufacture the monocrystalline and polycrystalline lies in the composition of the silicon substrate which is used to manufacture the PV cells and, in the end, the solar panels. As the name implies, polycrystalline point toward several numbers of crystals, on the other hand, monocrystalline point toward a single crystal.

As the size of the crystals are huge, the solar cell or the PV cell will be more effective, which explains the reason monocrystalline cells are near about 10% to 15%, when it is compared with the polycrystalline crystals.

2.4. USES OF SOLAR ENERGY

The capabilities of solar energy are massive, since near about 200,000times the total daily electric-production capacity of the world, is received by the earth with each passing day in the form of solar energy.

Nevertheless, solar energy itself is free for all, the high cost of its collection, conversion, as well as storage still restricts the usage of solar energy in several numbers of different places, all across the world.

The solar radiation or light coming from the Sun can be transformed either in the form of thermal energy (which is heat) or in the form of electrical energy (which is electric current); nevertheless, the former is much easier to achieve.

2.4.1. Thermal Energy

Out of the most common devices that are being used to harness solar energy and transform solar energy in the form of thermal energy or heat, are flat collectors, which are used for solar heating applications. The reason behind this is the intensity of the solar radiation at the surface of the earth is so low, these flat collectors must be large in area to fulfill the process more effectively.

Even in the regions of the world, where the solar radiation or sunlight found in plenty amount or sunny days, for example, a collector must have surface area of near about 40 square meters (which is nearly 430 square feet) to collect the sufficient amount of energy to fulfill the energy requirements of an individual.

The most extensive used flat-plate collectors are consisting of a blackened metal plate, that is covered with one sheet or two sheets of glass, that is heated with the help of solar radiations or sunlight falling on it.

This heat is then conveyed to the air or water, which are known as carrier fluids, that flow past the back of the plate. The heat can be used directly, or also, the heat can be transferred to the other medium of storage. Generally, flat-plate collectors are used for solar water heaters and house heating.

The storage of the heat for use at night or on cloudy days is generally achieved with the help of using the insulated tanks to store the water heater through the course of the sunny days. These kinds of systems can provide a home with hot water that is drawn from the storage tank, or, with the warmed water flowing across the tubes in floors and ceilings, it can provide space heating.

Generally, the flat-plat collectors are the heat carrier fluids to the temperature which is ranging from near about 66°C to 93°C (nearly 150°F to 200°F). The effectiveness of these sorts of collectors (for example, the amount of the energy received that they transformed in the form of usable energy) ranges from 20% to 80%, generally relies on the design of the collector.

Apart from this, the other method of the transformation of thermal energy is found in the solar ponds, which are the bodies of saltwater that are particularly designed to gather and store solar energy. The heat, which is extracted from these kinds of ponds, allows the generation of chemicals, chemicals, food, textiles, along with other industrial products and also, can be used to supply heat to the greenhouses, swimming pools, and livestock buildings as well.

Every so often, the solar ponds are used to generate the electric energy with the help of the use of the organic Rankine cycle engine, which is a comparatively effective and cost-efficient method of conversion of solar energy, which is particularly useful in remote areas. Solar ponds are fairly costly to install and maintain and are commonly restricted to warm rural regions.

Concerning the smaller scale, the energy coming from the Sun can also be used to cook in specially designed solar ovens. Conventionally, the solar ovens concentrate the sunlight from an extensive area to a central point, where a vessel that is consisting of a black surface helps to transform the solar radiations in the form of heat. Generally, the solar ovens are portable in size and necessitate no other kind of fuel inputs.

2.4.2. Electricity Generation

Solar radiations may be directly transformed in the form of electric current with the help of PV cells (also known as solar cells). With these kinds of cells, a small electric voltage is produced when the light reaches the junction between a metal and a semiconductor (for example, silicon) or the junction between the two various kinds of semiconductors. The energy, which is produced with the help of a single PV cell is commonly only about two watts.

With the help of connecting large numbers of individual cells together, nevertheless, just similar to the solar-panel arrays, hundreds or even thousands of kilowatts of electricity or electric energy can be produced in a solar electric plant or a large household array.

In the present interval of time, most of the energy effective PV cells is only near about 15% to 20%, and, since the intensity of the sunlight is low to start with, large, and expensive assemblies of these kinds of cells are needed to generate even moderate amounts of power.

Small PV cells that work on the solar radiations or light coming from the Sun, along with the help of artificial light have found major use in applications that necessitates low power, just similar to the case of power source for calculators and watches.

Larger units have been used to supply the power for water pumps and communications systems in remote regions and for weather and communications satellites.

Classic crystalline silicon panels and developing technologies with the help of thin-film solar cells, which is consisting of building-integrated PVs, can be installed with the help of the owners of the home and businesses on their rooftops to substitute or augment the traditional electric supply.

Concentrated solar power plants use concentrating, or focusing, collectors to focus the solar radiations or light from the Sun received from an extensive area onto a small blackened receiver, in this way, significantly elevating the intensity of the light to generate the high temperatures.

The sets of carefully aligned mirrors or lenses can concentrate a sufficient amount of sunlight to heat a target to the temperature level of near about 2000°C (or 3,600°F) or more.

This heat can then be utilized to operate the boiler, which, as an outcome, produces steam for a steam turbine electric generator power plant. To generate the steam directly, the movable mirrors can be aligned in such a way that they concentrate a huge amount of solar radiation on the blackened pipes with the help of which water is circulated and so as to heat.

2.4.3. Solar Water Heating

The application of solar energy extends to water heater systems. Most of the solar water heating solutions create hot water, which is utilized in the houses (Figure 2.4).

Figure 2.4: An illustration of solar water boiler.

Source: Image by Flickr.

Solar water heaters necessitate a rooftop cell to absorb the heat coming from the Sun and transfer that heat to the water tank. Generally, solar water heaters have 5 years to 10 years of payback, as per the United States Department of Energy (DOE).

Apart from this, the other application of solar energy, more particularly in the southern United States and the south-western United States, is heating swimming pools. Water is circulated to a collector where the water is being heated with the help of the sunlight and then pumped back into the pool.

With costs between 3000 dollars and 4000 dollars and a lifetime of near about 1.5 years to 7 years. According to the United States DOE, which states that "solar pool heating is the most cost-efficient application of solar energy in several kinds of different climates."

2.4.4. Solar Heating

The general application of solar space heating systems is consisting of powering radiant floors or pairing with a forced hot air (FHA) system to supply heat to a house. Passive solar home design can also heat the homes and businesses in the winter with the help of considering the placement of the windows and the selection of the materials that are being used in the construction.

2.4.5. Solar Ventilation

Solar ventilation solutions, for example, solar attic fans, can decrease the load of the HVAC (which is short form for heating ventilation and air conditioning) with the help of cooling the home through the course of summers. Solar ventilation can be a good option if an individual is not able to install a solar PV system that offsets the entire electricity use of the house. Apart from this, one innovative product is the Sola-tube solar attic fan (Figure 2.5).

Figure 2.5: An illustration of solar attic fan.

Source: Image by Pxhere.

Industrial applications as well as industrial applications of the solar process heat is consisting of solar ventilation technologies that can pre-heat the air of a building in cold climates, which decreases the energy costs.

2.4.6. Solar Lighting

In the present interval of time, solar lights have mostly been used at every place and can be found everywhere from home landscaping and security lights to the road signs and street lights. These kinds of solar lighting technologies for the homes are cost-efficient and readily available from basic to high-end designs everywhere from the local hardware store to e-commerce websites, for example, Amazon.com.

Apart from this, one innovative application of indoor solar lightning, feature on Mashable, in the Sola-tube skylight that adds natural light, along with the decreasing the use of energy.

2.4.7. Portable Solar

In the connected world, smartphones, and tablets are the necessities and, let us face it, every so often running low on the battery. Portable solar PV chargers can keep the personal devices that operate on electricity, charged on the go. The technology already exists to combine the solar cells or PV cells into smartphones and has been used in watches since the period of 1970s (Figure 2.6).

Figure 2.6: An illustration of portable solar charger.

Source: Image by wallpaper flare.

Taking an example of Japan, researchers have developed light-weight, water-resistant PV cells that could someday be sewed into the clothing to supply the power to the devices.

2.4.8. Solar Transportation

Vehicles that operate on solar power may be the future, with existing applications that are consisting of buses, trains, airplanes, and race cars built with the help of the students in Australia and the United States. A car, which is entirely powered with the help of solar cells or PV cells is even slated for commercial release in the year 2019.

This application of solar energy is not yet extensively available unless an individual or a person owns an electric car or electric vehicle (EV) and charges it with the help of the solar panels (commonly with the help of the solar-connected battery).

2.4.9. Wearable Solar Tech

In the present interval of time, powering the consumer's electronics has become a usual solar power application, for example, solar-powered chargers, such as Anker's Power port, can charge anything, ranging from a smartphone to a tablet or e-reader. There are even flashlights that are powered by solar energy, that can be charged with the help of just being exposed to the solar radiation of sunlight (Figure 2.7).

For those who are curious about the products in solar technology, here is the list of the top 10 application of solar devices, mentioned below:

- Cell phones;
- Wearables;
- Music speakers;
- Solar air conditioning;
- Tablets;
- Thermostats;
- Solar dryers;
- Solar visor radios;
- Freezers/mini-fridges; and
- Rechargeable flashlights.

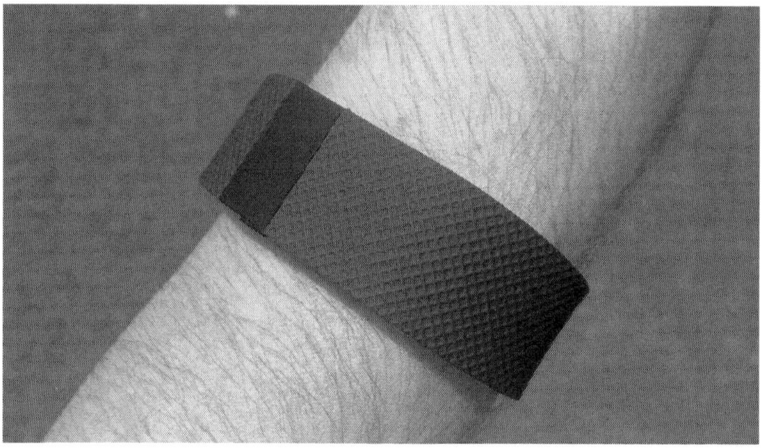

Figure 2.7: An illustration of wearable solar tech.

Source: Image by Wikipedia.

As the world moves away from the use of the fossil fuels and search for renewable sources to the daily basis electronics, there is no doubt that the solar energy will continue to adopt by the huge amount of people for any kind of product that can be exposed to solar radiations or the sunlight.

The main objective, utilization of the solar products to reinvent the methods and techniques that a person thinks of power usage, and reverse the expectation for how devices should be charged. And as the conversion efficiency of the solar endures to enhance, this trend will expand into the new markets as well as new products.

For instance, wearable solar could soon make Apple Watches, and Fitbits much more appealing to the consumers, not like their standard counterparts, smartwatches that are solar-powered do not require to be plugged in to recharge every night.

2.4.10. Other Applications

Solar energy is also utilized on a small scale for purposes other than those which are mentioned in the above topics. To some nations, for example, solar energy is used to generate salt from seawater, with the help of the process of evaporation.

In the same manner, solar-powered desalination units convert the saltwater into drinking water with the help of transforming the solar radiations in the form of heat, directly or indirectly, to conduct the process of desalination.

Solar technology has also developed for the clean as well as renewable production of hydrogen as a substitute source of energy. Imitating the process of photosynthesis, artificial leaves are silicon-based devices that utilize solar energy to separate the water in the form of hydrogen and oxygen, virtually leaving no pollutants.

In addition to this, further work is required to enhance the effectiveness as well as cost-effectiveness of these kinds of devices for industrial use as well.

2.5. COMPONENTS OF SOLAR SYSTEMS

Solar power is one of the cleanest, most reliable forms of renewable energy that is available in the present interval of time. Solar energy can be used in various kinds of forms to power houses or businesses.

Solar-powered PV panels transform the solar radiations or sunlight in the form of electric current with the help of exciting the electrons that are present in the silicon cells with the help of light that is coming from the Sun. This electric current can be utilized to supply renewable energy to the house.

To understand this process deeper, let us take a look at the components of the solar energy that compose an entire solar power system.

2.5.1. The Roof System

In most solar systems, the solar panels are situated on the roof. An ideal site will have no shade on the panel, more particularly through the course of the prime sunlight hours of 9 A.M. to 3 P.M. Generally, a south-facing installation will give the optimum potential for the solar power system, apart from this, other orientations or other settings may provide enough production of electricity (Figure 2.8).

Figure 2.8: An illustration of solar panels installed on the rooftop of a house.

Source: Image by Wikimedia commons.

Trees or other aspects that produce shading through the course of the day will generate a considerable amount of decrement in the production of power. The significance of the shading and efficacy cannot be overstated. To the solar panel, if even just one of its 36 cells is under the shading, the production of power will be decreased by more than half.

The well-known installation contractors, for example, NW Wind and solar, employ a device which is known as Solar Pathfinder to carefully determine the potential regions of shading before the process of installation.

Not every roof has the correct orientation or angle of inclination to have the benefit of solar radiation. That is why some kinds of systems are designed with pivoting panels that provide assistance to track the Sun in its journey across the sky.

Non-tracking PV systems should be inclined at an angle that is equal to the latitude of the site to absorb as much amount of energy throughout the year. Apart from this, alternate orientations and/or inclinations can be used to optimize the production of energy for particular times of the day for the particular seasons of the year.

2.5.2. Solar Panels

Solar panels are also known as modules. These modules are consisting of PV cells or solar cells. Solar cells are made up with the help of silicon that helps in converting the incoming solar radiations in the form of electricity, in spite of the heat. ("Photovoltaic" means electricity from light—photo = light, voltaic = electricity).

Solar PV cells are consisting of a positive as well as a negative membrane of silicon which is situated under a thin slice of glass. As the photon of the light coming from the Sun strikes on these solar PV cells, they knock off the electrons that are present in the silicon.

The negatively charged free electrons are specifically attracted to one side of the silicon cell, which produces an electric voltage that can be gathered and channeled. This electric current is collected with the help of wiring the individual solar panels together in sequence to form a solar PV array.

According to the size of the installation, several numbers of strings of the solar PV array cables end up in one electrical box. This electrical box is known as a fused array combiner (Figure 2.9).

Figure 2.9: An illustration of solar panels.

Source: Image by Air Force Civil Engineer Center.

Contained within the combiner box are fuses that are designed to safeguard module cable, along with the connections that deliver power or electric current to the inverter. The electric energy produced at this time is direct current (DC) and must be transformed in the form of alternating current (AC) which is suitable for use in homes or businesses.

2.5.3. Inverter

Conventionally, an inverter is situated in a location that is accessible, as close as practical to the modules. In a residential application, the inverter is every so often placed to the exterior side-wall of the home neat to the electrical main or sub-panels. As an inverter is quite noisy, this should be taken into the consideration, at the time of choosing the location of the inverter.

The work of an inverter is to turn the DC which is produced with the help of the solar panels into 120 volts AC, that can be put into immediate use with the help of linking the inverter directly to a dedicated circuit breaker in the electrical panel (Figure 2.10).

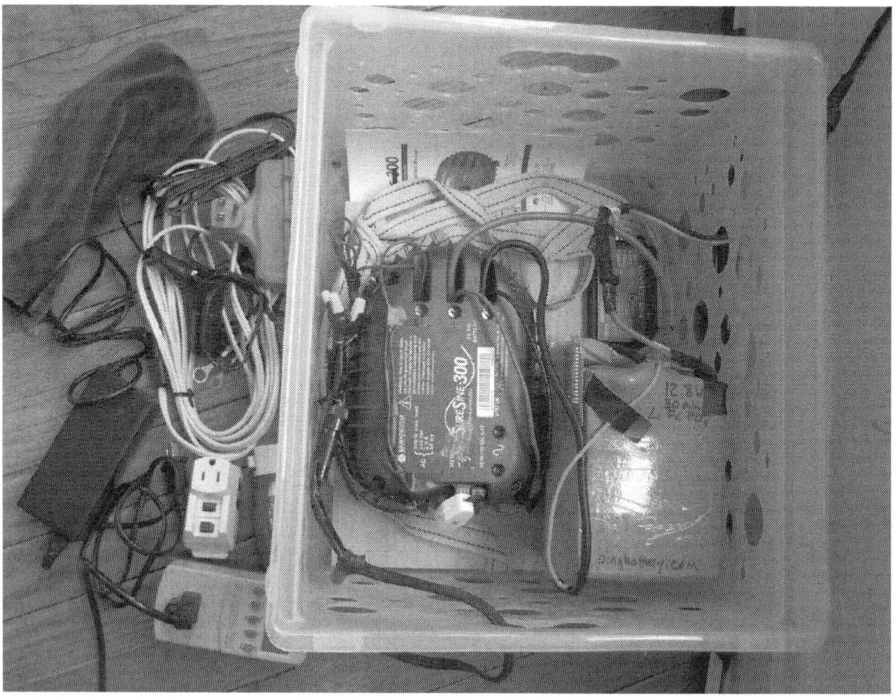

Figure 2.10: An illustration of solar inverter.

Source: Image by Flickr.

The inverter, electricity net meter, along electricity production meter are linked in such a way that the power that is being generated with the help of a solar electric system will first be consumed with the help of the electrical loads that are currently in operation.

The balance of the power of electric energy, which is generated with the help of the solar electric system, passes through the electrical panel and out onto the electric grid.

2.5.4. Net Meter

The solar electric system that is connected to the utility grid, the DC power from the solar array is then transformed into 120/240 volts AC power and fed directly into the utility power distribution system of the building.

The power is "net metered." In simple words, the meaning of net-metered is that it decreases the demand of the power from the utility when the solar array is producing the electricity, in this way, it decreases the utility bill.

These grid-tied systems are automatically turned off if the utility power goes offline, safeguarding the workers from the power being back-fed into the grid through the course of an outage. These sorts of electric systems that are driven with the help of solar systems are known as "on-grid" or "battery-less" and make up near about 98% of the solar power systems being installed in a present interval of time.

2.5.5. Other Benefits of Solar

With the help of decreasing the utility bill of a building, these kinds of systems not only pay for themselves over the interval of time, but in addition to this, they also aid in decreasing the air pollution that is being caused by the utility firms.

For instance, solar power systems assist to elevate something that is known as "peak load generating capacity," in this way, saving the utility from turning on expensive and polluting supplemental systems through the course of the peak demand.

The more locally produced solar electric power systems that are installed in a given utility's service area, the less capacity the utility requires to build, in this way, saving each and everybody from funding costly additional power generating the sources.

Contributing clean, green power with the help of solar electric systems assists to generate employment and is a great way to decrease pollution along with the other issues that are produced by the electricity that is being generated with the help of fossil fuels.

Electricity generating systems that are driven on solar energy provide assistance to decrease the effect on the environment, and also, it is cost-effective.

2.6. CONSIDERATIONS TO KEEP IN MIND BEFORE INSTALLING SOLAR PANELS

Before taking any step, to install a solar panel in the home, ensure that the solar power befits the home. Then make sure that the appropriate amount of light from the Sun or solar radiation is available at site. The suitability of solar energy fluctuates significantly with how much sun an area receives. If the area does not receive enough amount of solar radiation or sunlight, putting money in the solar panels will not be the idea to take.

After identifying that the area receives a sufficient amount of solar radiation or light from the Sun, ensure the installation space is available. If an individual has an open space to spare in the backyard, it would be a perfect site for ground mounts. The backyard option is ideal for those roofs that are heavily shaded or not structurally essential for the solar panels.

In addition to this, it is very important to acknowledge that the local laws for solar panel installation to avoid getting into trouble with the local authorities. An individual can get the information from their local solar energy consultant.

2.6.1. Environmental Impact

Nevertheless, solar energy is well thought out to be one of the best, cleanest as well as renewable source of energy between the available sources, it has some environmental effects as well. Solar energy employs PV cells or solar cells to generate solar power.

Although, manufacturing the PV cells to generate that energy necessitates silicon and produces some waste products. Inadequate regulation of these sorts of material can take into the direction of dangerous exposure to humans as well as the environment.

Installation of the solar power plant may necessitate a huge area of land, which can affect the existing ecosystem. Solar energy does not pollute the air when it is transformed in the form of electricity with the help of solar panels. Solar energy is found in plenty amount and in addition to this, solar energy does not help in global warming.

2.7. IMPORTANCE OF SOLAR ENERGY

Solar energy is the prime major renewable source. Solar energy is capable to fulfill several numbers of difficulties all across the globe. There are several numbers of various reasons to promote its share in the industry of energy. This power source is achieving more and more popularity, and the reason behind this is its versatility with numerous advantages to people as well as the environment.

As per the United States DOE, the amount of solar radiation or light coming from the Sun received by the earth in one hour is more than the total energy used in an entire year, all across the globe.

In the year 2015, solar energy was the fastest energy sector with an increment of near about 33%, as per the reports of Bloomberg. The environmental benefits are the main drivers to promote solar energy.

2.7.1. The Promise of Solar Energy

All life on the earth is supported with the help of the Sun. This amazing resource radiates a huge amount of energy and gives us both heat as well as light with the help of fusing the hydrogen into the helium at it once. This process is known as solar radiation. Only about half of this solar radiation makes it to the surface of the earth.

The rest of the solar radiation is either absorbed or reflected with the help of the clouds as well as the atmosphere. But still, enough amount of the power that is coming from the Sun is received by the earth, to fulfill the requirement of power of all mankind, which is millions of times over. Solar energy, which is the power coming from the Sun, and is vast, never-ending, and a clean source.

Solar radiation, or solar energy, can be used directly to heat or lightning the homes and businesses, for producing electricity, and for hot water heating, solar cooling, and several kinds of other commercials as well as industrial uses.

Most critical, given the increasing concern over the sudden changes in the climate, is the fact that the production of electricity with the help of solar radiation represents a clean substitute to the electric energy from the fossil fuels, no global warming pollution, no risks of the electricity price hike, and no dangers to the human health.

The solar resource is massive. As per the United States DOE, the amount of sunlight or the solar radiation that reaches the surface of the earth in an hour and a half is sufficient to control the consumption of the energy of the entire world for a year. Just 18 days of sunshine on the earth is consisting of the same amount of energy as is stored in all of the reserves of natural gas, coal, and oil of the planet.

And, once a system is in place to capture the solar resource and transform it into the form of useful energy, the fuel is free.

2.7.2. The Growth of Solar Energy

From the year 2008, the U.S. installations have increased 17 times from the value of 1.2 gigawatts (GW) to an estimated 30 GW till the present interval of time. It is a sufficient amount of power that can power near about 5.7 million average American homes.

From the period of 2010, the average cost of solar PV panels has decreased by over 60%, and the cost of a solar electric system has dropped by near about 50%.

Solar power is now well thought out as economically competitive with the traditional energy sources in the various kinds of states, which is consisting of California, Hawaii, Texas, and Minnesota.

2.7.3. The Basics of Solar Energy

What are the basics ofsolar energy systems? First of all, solar energy systems fluctuate according to size as well as the usage. Residential systems are found on rooftops all across the United States, and businesses are starting to install solar panels to reduce or lessen their utility bills or energy costs.

Utilities are also building large solar power plants to provide cleaner energy to all customers that are linked to the grid.

Irrespective of the specifics of a given installation, there are two main kinds of solar energy technologies, which are PV and concentrating solar power (CSP). Most people are familiar with the PV technology, and the reason behind this is that the solar panels they see more and more on the tops of the buildings or ones placed on the International Space Station.

When the Sun shines onto one of those solar panels, photons from the solar radiation or the light coming from the Sun are absorbed with the help of cells that are present in the panel, which produced an electric field across the layers and causes an electric current to flow. The installation of the PVs may be ground-mounted, rooftop-mounted and wall-mounted as well.

The PVs may be mounted in a permanent orientation to maximize the generation of the electricity and value, or they may be mounted on trackers that follow the Sun across the sky. Rooftops PV panels make solar power viable in virtually each part of the United States.

In a sunny location, for example, Los Angeles or Phoenix, a five kilowatts residential system generates near about 7000 kilowatts to 8000 kilowatts with each passing day, which is approximately equal to the electricity usage of a typical United States household.

The second technology is focusing on solar power (which is also known as concentrated solar thermal and CSP). Concentrated solar thermal is mainly used in very large power plants and is not suitable for residential use.

This technology employs mirrors to reflect and focus the solar radiation on to the receivers that gather the solar energy and then transform it into the form of heat. The heat is then used to drive the heat engine, which is generally a steam turbine, that is linked with an electric power generator which is then used to generate the electric energy or electric current.

2.8. ADVANTAGES OF SOLAR ENERGY

2.8.1. Renewable Energy Source

Out of all the advantages of solar panels, the most necessary thing is that solar energy is a truly renewable source of energy. Solar energy can be harnessed in all regions of the world and also it is available each day.

Solar energy is a never-ending source of energy, it is not like some of the other sources of energy. Solar energy will be accessible as long as Sun exists, and in this way, the solar radiation will be available for every person for at least 5 billion years when according to scientists, the Sun is going to die.

2.8.2. Reduces Electricity Bills

Since people will be fulfilling some of their energy requirements with the help of the electric power that has been produced with the help of the solar system, their utility bills will drop. How much a person saves on his bill will rely on the size of the solar system as well as the usage of the heat or the electricity. In addition to this, not only will a person will be saving on the electricity bill, there is also a probability to receive payments for the surplus energy that an individual export back to the grid. If a person produces more electric energy as compared to the use (considering that the solar panel of that person is linked to the grid).

2.8.3. Diverse Applications

The energy produced with the help of the solar system can be used for several kinds of various purposes. A person can produce electricity (with the help of PVs) or they can produce heat (solar thermal). Solar energy can be used to generate electricity in areas without access to the energy grid, to distill water in regions with limited clean water supplies, and to power the satellites that are present in space.

Solar energy can be combined into the materials that are used for buildings. Not long ago Sharp introduced transparent solar energy windows.

2.8.4. Low Maintenance Costs

Generally, solar energy systems do not necessitate a lot of maintenance. A person only requires to keep the solar energy systems relatively clean, in this way, cleaning them a couple of times per year will do the job. If in doubt, a person can always rely on specialized cleaning organizations.

Most dependable solar panel manufacturers offer a warranty of near about 20 years to 25 years. Also, there are no moving parts, there is no wear and tear. The inverter is only the part, out of all, which requires to be changed after the period of near about 5 years to 10 years, and the reason behind this is that inverter is working non-stop to transform the solar energy in the form of electric energy and heat (solar PV versus solar thermal).

Apart from the inverter, the cables also require to be maintained to make the power system runs at maximum efficacy. In this way, after transforming the initial cost of the solar system, a person can expect very little spending on maintenance and repair work.

2.8.5. Technology Development

Technology in the solar power industry is consistently improving, and enhancements will be increased in the coming interval of time. Innovations in quantum physics and nanotechnology can potentially elevate the efficacy of the solar panels and double, or even three times, the electrical input of the solar power systems.

2.9. DISADVANTAGES OF SOLAR ENERGY

2.9.1. Cost

The initial cost of purchasing a solar system is fairly high. This is consisting of paying for the solar panels, inverters, batteries, wiring, as well as for the installation. Nonetheless, solar technologies are continuously developing, in this way, it is safe to assume that the cost will decrease in the coming interval of time.

2.9.2. Weather Dependent

Nevertheless, the solar energy can still be gathered through the course of cloudy as well as rainy days, the efficacy of the solar system drops. Solar panels rely on the solar radiations or light coming from the Sun to efficiently collect solar energy. In this way, a few cloudy, rainy days can have a considerable impact on the energy system.

A person should also take into consideration that solar energy cannot be gathered through the course of the night. On the contrary, if a person also needs his or her heating solution to work at night or through the course of the wintertime, thermodynamic panels are an alternative to consider.

2.9.3. Solar Energy Storage Is Expensive

Solar energy has to be used right away, or the solar energy can be stored in the batteries. These batteries, used in off-the-grid solar systems, can be charged through the course of the day, in such a way that the energy can be used at the night. This is a good solution for utilizing the energy all day long but in addition to this, it is also quite expensive.

In most instances, it is smarter to just use solar energy through the course of the day and take energy from the grid through the course of the night (a person can only do this if the system is linked to the grid). Luckily, the requirement of the energy is generally higher through the course of the day, in this way, a person can fulfill his or her requirement of the energy with the help of solar energy.

2.9.4. Uses a Lot of Space

The more electricity a person desire to generate, the more numbers of solar panels he will need, as he or she wants to gather much solar radiation as possible. Solar PV panels necessitate a lot of space, and some roofs are not big enough to fit the number of solar panels that he or she would like to have.

2.9.5. Associated with Pollution

Nevertheless, the pollution that is related to the solar energy system is far less when compared with the other sources of energy, solar energy can be linked with pollution. The transportation as well as the installation of the solar system has been linked with the emission of GHG (greenhouse gases).

2.10. CONCLUSION

Solar power or solar energy is a massive source of directly useable energy, and with the help of solar radiations, several numbers of various devices can be a drive or powered and can be charged.

Most of the surface of the earth receives enough amount of solar energy to permit low-grade heating of water and buildings, nevertheless, there are a huge amount of variations with the latitude and season. At low latitudes, simple mirror devices can focus solar energy sufficiently for cooking and also, to power the steam turbines.

The energy of light shifts electrons in some of the semiconducting materials. This PV effect is having the potential of producing electricity on a large scale. Nevertheless, the present low efficacy of the solar PV cells requires very large areas to supply the electricity demands.

Direct use of solar energy or solar power is the only renewable means that is having the potential of ultimately supplanting the existing global energy supply from nonrenewable source, but at the expense of a land area of at least half a million kilometers.

REFERENCES

1. Ashok, S., (2020). *Solar Energy | Description, Uses, and Facts*. [Online] Encyclopedia Britannica. Available at: https://www.britannica.com/science/solar-energy (accessed on 5 May 2021).
2. Dana, O., (2019). *Why the Future of Solar is Bright-Renewable Energy World*. [Online] Renewable Energy World. Available at: https://www.renewableenergyworld.com/2019/11/12/why-the-future-of-solar-is-bright/#gref (accessed on 5 May 2021).
3. Freedom Solar Power, (2018). *Seven Uses of Solar Energy | Freedom Solar Power.* [Online] Freedomsolarpower.com. Available at: https://freedomsolarpower.com/blog/7-uses-of-solar-energy (accessed on 5 May 2021).
4. Green Match, (2020). *Advantages and Disadvantages of Solar Energy (2020) | Green Match.* [Online] Greenmatch.co.uk. Available at: https://www.greenmatch.co.uk/blog/2014/08/5-advantages-and-5-disadvantages-of-solar-energy (accessed on 5 May 2021).
5. Marsh, J., (2020). *What is Solar Energy? Solar Power Explained | EnergySage.* [Online] Solar News. Available at: https://news.energysage.com/what-is-solar-energy/ (accessed on 5 May 2021).
6. PV Education, (2019). *Solar Energy | PV Education.* [Online] Pveducation.org. Available at: https://www.pveducation.org/pvcdrom/introduction/solar-energy (accessed on 5 May 2021).
7. Richardson, L., (2018). *The 5 Most Common Uses of Solar Energy in 2020 | EnergySage*. [Online] Solar News. Available at: https://news.energysage.com/most-common-solar-energy-uses/ (accessed on 5 May 2021).
8. Rinkesh, (2020). *What is Solar Energy and How Solar Energy Is Converted into Electricity-Conserve Energy Future*. [Online] Conserve Energy Future. Available at: https://www.conserve-energy-future.com/solarenergy.php (accessed on 5 May 2021).
9. Solar Electric Power Company, (2012). *Solar Power Advantages and Disadvantages*. [Online] Sepco-solarlighting.com. Available at: https://www.sepco-solarlighting.com/blog/bid/115086/solar-power-advantages-and-disadvantages (accessed on 5 May 2021).
10. Solar, N., (2015). *How Do Solar Systems Produce Energy?* [Online] NW Wind & Solar. Available at: https://www.nwwindandsolar.com/

solar-power-in-seattle-and-the-northwest/how-do-solar-systems-produce-energy/ (accessed on 5 May 2021).

11. Venture Solar, (2017). *The Importance of Solar Energy | Venture Solar*. [Online] Venture Solar. Available at: https://venturesolar.com/blog/the-importance-of-solar-energy/ (accessed on 5 May 2021).
12. Vijayalaxmi, K., (2020). *Why is Solar Energy Important?* [Online] Greenliving.lovetoknow.com. Available at: https://greenliving.lovetoknow.com/Why_Is_Solar_Energy_Important (accessed on 5 May 2021).

Chapter 3

Potential Use of Solar Energy in Agriculture Sector

CONTENTS

Solar energy acts as a boon for the agriculture sector. In the chapter, the potential use of solar energy in the agriculture sector, some of the solar energy technologies that are widely used in the agriculture sector are discussed. The chapter also recognizes several benefits of using solar energy in the agriculture sector. The chapter also explains the role of solar power irrigation as a solution to water management in agriculture. The concept of green technology is also explained in this chapter.

It also shed some light on solar-powered system micro-irrigation systems for agriculture. It also highlighted portable solar photovoltaic (SPV) powered pumping system for micro-irrigation system in vegetable cultivation. In the end, the chapter explains future trends in solar-powered irrigation systems (SPIS).

3.1. INTRODUCTION

It is generally seen that the energy sector plays a very critical role in the overall development of the economy of a country. In the current scenario, almost 85–90% of the world's primary energy comes from the burning of fossil fuels.

There is restricted storage of fossil fuels and one of the imperative reasons for the recession in the world's economy is the consistent rise in the price of these fossil fuels. To address the problem of this falling economy, and the energy sector's related issues, most countries are focusing on effective utilization of renewable sources of energy such as solar, thermal, wind, and hydro. Supplies of fossil energy became available around 200 years ago.

In addition, shortages of cropland, fossil energy (irrigation and fertilizers), freshwater, and biological resources now impacting agricultural production in several countries of the world. Although, fossil energy resources have started declining and this trend has been intensified after the year 2000.

The Use of renewable energy in the farming systems such as solar energy has several different practical applicability. Renewable energy can be taken into use for a variety of farm or agriculture activities that are usually done with electricity and petroleum products such as pumping water for irrigation, for domestic use, or for keeping livestock; powering processing operations, lighting farm buildings; and other uses.

These forms of renewable energy comprise solar energy, water, and wind power, wood from sustainable sources, oil from plants, other forms of biomass (plant material), and biogas (gas produced from the fermentation of crop and manure residues).

The overall agriculture yields dependent on the potential of solar plants to Convery solar energy into electrical energy and stored it into batteries to use later for different purposes. Solar energy is the most adequate option in comparison to other renewable energy sources because the solar energy level is in line with the air condition demand. In addition, solar energy technologies have a long history.

Between 1860 and the First World War, a variety of technologies developed to create steam by capturing the Sun's heat to run irrigation pumps and engines. Taking into consideration the significance of solar energy and the continuous increase in interest among people to use solar power for their daily usage, it becomes even more important to learn different usage of solar energy in the agriculture sector. In this chapter, the main area of focus will be on how one can maximize the use of solar energy for different usage in the agriculture sector.

3.2. SOLAR ENERGY TECHNOLOGIES

There are two methods to transform solar energy into electrical energy; a system using photovoltaic (PV) technology and the other one that uses solar capture heating systems. In the PV system, the sun rays are directly transformed into electricity with the help of semiconductors. Furthermore, in the method of heating, electrical power is generated via the thermodynamic processes, with the help of heat exchange equipment, which can be transformed into mechanical energy.

These two methods are centralized and non-centralized. It is often seen that the PV method requires heavy investment and therefore farmer resists to employ this technology. Although, in recent years, with advancements in technology of solar energy, thermal methods are used for power supply.

3.2.1. Photovoltaic (PV) Technology

Solar PV cells were invented at Bell Labs in the U.S. in the year 1954, and they are intended to use in space satellites for generating electricity since the late 1950s. In this type of technology, solar rays gathered via small plates which are semiconductor PV, are transformed into electricity. There are basically two ways in which PV cells can be built: flat panel and concentrator.

Solar cells are the most common flat panels where the light is directly brought to semiconductors and is transformed into electricity. Yet, in the concentrator cells, first the sunlight is amassed with the help of a reflector, concentrated, and then the solar cell connects. The solar cells are connected by solar modules. Power cells and solar modules have the potential to charge the battery and to form a system with output considerably needs that modules that work together and at the same time.

As it is well known that solar cells are linked together and making modules, also the modules for generating the adequate voltage and current, are fixed in series and parallel to that unit made in this way is known as the solar array.

3.2.2. Solar Dryer

It is usually seen that preservation of food through drying is one of the traditional and the most prevalent methods that can be used to improve the strength of the food. Drying food helps in eliminating the moisture, allowing food to be stored for a longer duration of time and be protected against corruption.

By lessening the microbial enzyme action and decreasing the chemical reaction speed, drying enhances the overall shelf life of the product. Furthermore, reducing the volume and weight of packaging and materials facilitates storage as well as transportation of products and lessens the whole process cost of conveying the product from producer to consumer.

In the case of drying, apart from averting the loss, the marketing can be controlled at the time of high season and potatoes required by many consumers (such as restaurants, barracks, etc.), can be conveyed in a dried form.

Using the Sun for drying the grain and crops is one of the traditional and widely adopted applications of solar energy. Solar dryers protect fruits and grain and vegetables, reduce losses, dry faster and more uniformly, and produce a better-quality product than open-air methods. Solar-drying technology offers an alternative, which can process the vegetables and fruits in sanitary conditions to international and national standards and with zero energy costs.

It saves energy, occupies less area, time, improves the quality of product, makes the process more effective and efficient and protects the environment. Ekechukwu et al. and Fuller (Fuller, 1995) have appraised many solar dryers, and compared their applicability and performance in rural areas.

Sharma has demonstrated a complete review of the various designs, details of operational and construction principles of the wide diversity of practically achieved designs of solar-energy drying systems, and a systematic approach for the categorization of solar-energy dryers has been progressed.

Solar dryers are traditional dryers so that additional equipment is added to enable a substantial fraction of the thermal energy needed for drying to be swapped by solar energy. In these types of the dryer, an organized, and usually enhanced drying process can be attained to attain good economic performance and high-class product quality.

Any impact of the weather conditions on the quality of product and the functioning of the dryer can be eradicated by using an independent energy source if required, and proper control facilities. The formation of the solar-assisted dryers is comparatively intricate in comparison to other dryers. They usually comprise a solar collector, a heat storage system, a fan, a heater/burner, and a control system. They have the potential to manage large quantities and offer a high quality product.

3.2.3. Ventilation Systems for Agricultural Applications

Therefore, the application of PV is one of the best choices because at the time of electricity breakdown, it can be taken into use to supply power, hence possibly saving thousands of birds. Direct current (DC) motors can also be operated directly with PV power and reduce the need for a highly expensive inverter.

Several governments of different countries provide subsidies and other incentives to farmers that make use of solar energy to produce electricity, particularly for pig farms, to appeal to other farmers to construct solar farms. It is also important to note that PV offers additional benefits, such as security of supply, and environmental and economic benefits over grid electricity supply and traditional energy sources.

3.2.4. Use of Solar Energy in Farm Houses

It is generally seen that there are some locations in villages in which it is extremely challenging to supply electricity and thus requires huge costs to install electricity towers. In such places, solar energy can work best as it can be accessed from anywhere and can provide the fuel and electricity required, for a home can provide home lighting lamps and other appliances using the energy used.

But the question remain persists for cooking and heating water heaters: What can be the possible solution for this? It is true to fix such problems by using water heaters and solar cookers.

3.2.5. Solar Home

Seamlessly PV building that has been worn and has become one of the most popular homes. And, it has the highest applicability in areas where there is no electrical grid. PV panels can be employed on the rooftop or walls to access sunlight. Solar energy is produced via PV panels at the same location, and extra energy can be stored for future usage.

This technology is inexpensive and assuring, and more power can be generated for the home. For areas, where electricity grids are not very useful, it helps the region supply with the least electricity problems. For example, a mountain or other terrain areas where supply electricity power is not possible, solar homes can be built. These homes are self-independent in generating electricity and can in fact supply power to other appliances installed at home.

3.2.6. Solar Cooker

Solar cookers are one of the best ways to make use of clean energy, which is available in abundance and is free to use for cooking food. A solar cooker occupied radiant heat by rays concentrating it on a surface or gathering it in a greenhouse space. Rather it should be used for cooking.

As it is well known that in developed countries about 90% of household energy consumption is devoted to cooking. These systems work almost in the same manner as an electric rice cooker or gas stove, such as brewing, roasting, cooking, reheating, or frying refrigerated cool foods, including cases that are probable with the solar cookers, hence there are two forms of solar cookers, direct, and indirect.

The direct solar cooker comprises an insulated box having a transparent window, giving entry to sunlight. It is based on the most advanced technology and is widely used for home cooking. In the indirect stove, fluid is used for the transfer of heat from the collector to a baking sheet.

3.2.7. Solar Water Heater Systems

Water heaters are one of the most common usages of solar energy that has been used extensively in both homes as well as in industries. Water heating systems are also available in natural convection and forced convection schemes.

Solar water heaters are often categorized into two parts: direct and indirect. In direct water heaters, water is gathered in the current collector, heated, and then conveyed for usage, but in the indirect water heater, the water amassed is used for heating a fluid. The limitation of direct solar water heaters is that after some time, the collector is jammed with a crime. Also, the thermal energy that is being conveyed to the water consumed is wasted.

In general, in solar water heaters, when there is no much usage of hot water, then it is stored in the reservoir so that it can be used later. In some types of solar water heaters, a water tank is employed just above the collector, and hot water for natural convection is stored in the tank.

In the indirect solar water heaters, a tank is fitted in a distinct area in a way to supply power to flow in the collector by the used pump. Because of that, this water heater is said to be under pressure.

3.2.8. Solar Greenhouse

The greenhouse is referring to a structure that is being used nowadays in the agriculture sector to grow plants with the finest quality. Currently, solar energy is taken into use for heating greenhouses, due to this reason, it is known as solar greenhouses, and solar energy is used to supply light to the greenhouse. One of the best things about this is that it reduces its dependency on fossil fuel, which would be used otherwise for heating purposes.

After the oil crisis in the year 1973, researchers were looking for a simple and convenient solution for the use of solar energy in the greenhouse. Overall, there are three ways in which solar energy can be used in greenhouses.

The first type is famous as the inactive greenhouse (Passive), and it is taken into use thermal energy from radiation solar heated greenhouse. In the greenhouse, the construction method has been adopted to ensure the maximum usage of solar radiation during the daytime and lead to the lowest energy losses during the night in greenhouses.

The second category of the greenhouse is known as greenhouse active (Active), and it uses the pickers and transfers heated fluid, in heating greenhouses. In the third form of greenhouses, PV cells are taken into use to transform the solar radiation energy into electrical energy so that it can be used in greenhouses.

3.3. ADVANTAGES OF SOLAR ENERGY FOR AGRICULTURE

It is well known that solar energy is creating its space in every place, whether it is home or office space. However, it also plays a prominent role in the agriculture sector, because it is the only sector that is far behind other sectors in terms of development.

Providing solar energy to agriculture is extremely important as it makes the sector independent and reduces its reliance on the power grid for consistent electric supply. Aside from lowering the electric bill, solar power allows farmlands to be much eco-friendlier.

In addition, it also offers productive solution to this that face serious challenge because of drought. It is generally seen that many farmers face huge losses because of this, but now is the time to continue pushing forward. Employment of solar into farm has a lot of benefits and it is discussed in subsections.

3.3.1. Cheaper Than Fossil Fuels

On many large farms, solar energy is comparatively less expensive than fossil fuel. It is because of a variety of reasons. During the initial phase of installation of solar energy, it is quite an expensive process. As solar energy PV cell demand huge investment, but it will pay off later.

It is just a one-time investment that saves significantly on electricity bills. In addition to this, the government of several countries also provides huge incentives, subsidies, and other benefits to those who employ solar power on their farm.

Apart from cutting down the electricity bills, solar energy is also a renewable source of energy. As the electricity cost is increasing continuously and electric companies usually charge more during the daytime, or when it is the hottest, with solar energy, one can save this grueling cost while maintaining the farm. In addition, solar energy frees farmers from electricity spikes or times when electricity is all consumed at once. However, on the other hand, electric companies will bill farmers much more for 'disrupting' the power system.

3.3.2. Sensible Solution to Drought-Related Problems

There are several areas in the world, where the problem of drought is prevalent. Many are finding a safe haven with the use of solar energy. Solar energy is currently one of the best options available during this relentless drought because it needs marginal water delivery.

For many, the overall productivity of agriculture has been reducing considerably because of the restricted water supply. Solar energy allows for the growth of PV panels as opposed to crops. While solar energy does not harvest fairly the similar yield, for affected land, it may be a good solution.

3.3.3. Not a Permanent Switch

One of the major misconceptions of solar energy is that it is a perpetual decision. But, in the majority of the cases, that is not simply true. Farmers can use solar energy as long as they require, and they can shift to electricity easily if in the future they need to do so. After the tenure, future crops will not be impacted by using renewable energy earlier.

Some are not concerned about converting permanently but are seriously impacted by the drought. Using solar energy can significantly assist farmers in sustaining drought conditions. Then, if they decide it is not for them, they are able to go back.

3.3.4. Potential Changes in Law

According to the government of several countries, they are focusing on reducing the dependency on fossil fuel as such as they can to sustain the crucial resources for future generations and to reduce climate change.

For this, some have set the target to entirely shift to solar and other renewable resources till 2030. Some have changed the law and require farmers to use solar energy in their farms instead of the power grid. To ensure wider acceptance of solar panel, several governments have started to give incentives and subsidies to those farmers that install solar panels into their farms.

As the world is becoming more aware of the impact of the burning of fossil fuel on climate change, it is more likely that there will be high pressure to make use of solar panels as much as possible.

Governor Jerry Brown believes that the transition away from fossil fuel energy would be a challenging task. For years, it has been the most economically and profitable wise solution. But this step will certainly help lead to a decarbonized future, according to Brown. It is believed that this will eventually serve as a model for other states and nations to follow.

3.3.5. Solar Energy Is Currently on the Rise

It is no doubt that within the past few years, there has been a significant rise in the use of renewable energy. People nowadays are much more aware and concern about their environment that forces them to use clean energy. In the future, this trend is expected to grow further. For farmers, this means that the cost to install solar energy is likely to decline.

According to the solar energy industries association (SEIA), has grown 60% in the last decade. However, the cost has plunged over 70%. With the growing popularity of renewable energy, prices are likely to fall in the near future. The advantages of solar energy will be seen through price reductions in upcoming years.

3.3.6. Becoming More Advanced

Like everything in our tech-savvy world, solar panels have developed significantly from their initial state. Now, the majority of the solar panels do not possess the glass substrate that the conventional models held. They are still attachable to walls and roofs. In addition, panels of the current era are much more environmentally friendly now. It is because of energy being required for its manufacturing process. The suitability of installation is one of the major benefits of solar energy.

3.3.7. Solar Storage Is Also Becoming More Affordable

In the past several years, it was observed that storing solar energy is one of the most expensive tasks. However, recently with the advancement in technology, it becomes much more economical because of economies of scale.

So far, battery storage has been increasingly widespread for the use of solar energy. Many people are making use of are lithium-ion batteries to preserve energy. Lithium-ion has the potential to store a large amount of energy in a limited space. But lithium-ion resources are limited in the world, therefore it is expensive.

The price to store renewable energy dropped aggressively in the last 2 years. As more companies have started extracting Lithium-ion resources from the earth, its cost is expected to drop 20 to 30% in upcoming years. Energy storage was highly imperative in areas that are vulnerable to natural disasters. In addition, storage of energy is also helpful for future crises.

3.3.8. Soil Becomes Better Habitat

It is generally seen that ensuring solar power energy creates a more habitat-friendly environment for wildlife and plants. Solar panels are usually surrounded by grasslands. The installation of the solar panel does not pose any threat or harm to plants. In addition, it can also be maintained easily through hedging. Moreover, animals are able to have more room to graze through, creating a pasture-like setting. They will not be damaged by the above or underground panels.

3.3.9. Most Efficient Way to Halt Global Warming

Within the past few years, global warming has been one of the most widely discussed topics. Making use of the renewable source of energy as much as possible would be significantly helpful in reducing a carbon footprint. As it is well known that solar panels produce renewable energy, one can use it to its maximum power without having any impact on the environment.

Not only that, but solar farms are able to disperse large amounts of clean energy. This is because of the excess power being distributed onto main grids. It is often seen that in many countries, especially developing ones, coal-based electric generation is the main source of energy that contributes significantly to global warming.

Therefore, the implementation of solar energy becomes even more important for developing countries. It would be possible only if there would be pressure on developing countries on international platforms. Thus, doing all such efforts would be significantly helpful in reducing global warming.

3.4. CONCEPT OF GREEN TECHNOLOGY

The green technology is a broader term that offers more environmentally friendly solutions. Green technology is believed to be one of the elite possible ways that can help in lessening the environmental damages caused by the technologies and the products designed for making the life of people easier.

Because of this, it has crafted its image as an environment healing technology that contributed significantly in maintaining the environment as well as the ecosystem. It is extensively believed by the masses that green technology plays an eminent role in boosting the overall profitability of farmers while at the same time preserving natural resources and diminishing environmental degradation.

The risk associated with poorly designed technologies such as petroleum products is making the existing situation even worse.

The "clean" technologies, on the other hand, are well known for offering low risk alternatives. Green technology covers a broad group of materials and methods for generating energy to non-toxic cleaning products.

The only aim behind this approach have been severely vital as people expect a dramatic innovation that can progress their conditions and making their life easier.

The development of alternative technology such as green technology is mainly concerned with providing benefits to the entire planet by truly protecting the environment. Although it is highly challenging to exactly outline the areas that come under the category of green technology, yet green technology plays a crucial role in tackling the emerging issues of sustainability because of the recent advancement in science and technology.

This technology should possess the ability to address the present needs of society in ways that can offer a sustainable and secure future without depleting or damaging natural resources. In short, green technology can be defined as the technology that helps in achieving the needs and wants of the existing generation while at the same time assessing the need of future generations by preserving the nonrenewable resources for them.

In terms of technology, products can be fully re-used or reclaimed by efficiently minimized pollution and waste in the cradle to the grave cycle of the manufacturing process. The recent advancement in technology demands the need for the production of alternative fuels as a new means of energy efficiency and producing energy.

The execution of such innovative technology is eminent for the agricultural sector to ensure the consistent food supply for the gradually increasing population and mitigating the present-day's concern of climate change and global warming.

Irrigation is believed to be one of the most energy-intensive activities in the farming sector. The significance of irrigation of small-holdings is expected to surge in the coming decades, especially in the developing countries, because of the steady rising population and because the majority of land-holdings are small, especially in Africa and Asia.

Hence, one of the viable options for deriving power through green energy technology is from biomass, solar, and wind energy.

PV pumping systems are believed to be appropriate for energy and water-saving methods of irrigation such as sprinkler and drip irrigation because of their advantage in the low-energy range.

The significant rise in fuel price and unreliable supply of electricity is at present the major roadblock for achieving assured irrigation for crop cultivation.

Solar pump therefore is considered to be the best alternative in comparison to the present electric motor operated pumps for irrigated crop production primarily for the off-grid rural areas.

3.4.1. Solar Photovoltaic (SPV) System

It is generally seen that electrical energy plays the central role of all developmental efforts in both the developed as well as developing nations because conventional energy sources are limited in nature and are draining at a higher rate.

In the last few decades, energy-related problems are increasing consistently and are appealing to the great interest among environment experts as it comprises the perfect use of resources, responsible for the environmental impact due to the emission of pollutants and the consumption of conventional energy resources.

Direct conversion of solar energy to electricity can be done conventionally through the use of PV cells, which make use of the PV effect. PV effect depends on the engagement of photons, with energy equal to or more than the band-gap of PV materials. In this, some loss can be averted easily as it is probable to avoid band-gap limitations by cascading semiconductors of different band-gaps.

PV modules have the potential to produce electricity directly from light without noise, emissions, or vibration. Sunlight is free, can be used by anyone, but power generation cost is markedly high, although prices are steadily plunging. The density of solar energy is comparatively low as PV modules generally require a larger area for small amounts of energy generation.

The major component in grid-connected PV systems is the inverter that transforms DC power generated by PV array into AC power consistent with the power quality and voltage requirement of the utility gird. Silicon solar cells are perhaps considered the simplest and most extensively used for terrestrial and space applications.

The PV system can be perceived as a trusted source of electricity production that helps in lessening carbon emission and energy resource-saving, even if existing technologies are applied. Further, the development of effective and efficient solar cells, the type, and proportion of material used in the solar cell, and the system design for greater use of recycled material will reduce the greenhouse gas emissions and energy requirement. Net annual carbon emission mitigation potential from 1.8 kW solar PV pump at average solar radiation of 5.5 $kWh/m^2/day$ is about 2085 kg from diesel-operated pumps and about 1860 kg from petrol operated pumps. The carbon dioxide emissions diminution potential is greater in the case of diesel substitution as compared to petrol substitution. It is chiefly because of the low efficiency of fuel utilization in the diesel engine pump.

The electricity demand for irrigation activity is continuously increasing, as the price of an electric-powered pump is relatively low to a diesel engine driven pump. The solar pump can be used as an alternative for small scale irrigated crop production in the off-grid areas of the tropical countries supplied with a copious supply of solar energy.

The solar radiation varies from 4.0 to 6.5 $kWh/m^2/day$ and the bright sunshine hours vary from 6 to 9 h/day. There is a larger area that needs to be irrigated where most of the areas have no grid connection. Solar PV pumps can play an eminent role in irrigating these lands for adequate production of agriculture material and to intensifies cropping intensity.

In regions that have high insolation levels, PV pumping systems were technically appropriate for use, environment friendly and also economical to use in comparison to diesel engine driven pumps.

The energy crisis is severe during the time of irrigation season that can be a threat to the economic development of any country. Ensuring far and wide use of solar pumps, the government of countries can intensely lessen the energy demand needed for the operation of irrigation systems.

Though the initial cost acquiring and implementation of a solar pump is comparatively greater than a conventional diesel engine operated pump, a solar pump has lower maintenance and repair cost, which means over a longer duration period, it is more cost-effective in comparison to a conventional pump. In addition, a solar pump is pollution free and environment friendly water pumping system.

3.4.2. Solar Water Pumping

It is worth noticing that nowadays, electrical and diesel-powered water pumping systems have been extensively used for irrigation applications. The gradual depletion of conventional energy sources and their subsequent environmental impacts have put an alarming situation across the world that results in forcing people to move towards picking solar photo-voltaic pumping system in a sustainable manner.

In some parts of the world, especially in developing countries, farmers are still using hand pumps for irrigation of small patchs of land.

Electric and diesel operated have been used mostly for lifting water from bore well, dug well, and other irrigation systems. Lifting of water by hand pump needs significant effort and is considered as one of the most intricate tasks.

In a similar manner, non-availability and restricted supply of grid-connected electricity in the rural areas forced the people to make use of diesel to carry out their day-to-day farming activities. All this necessitates the need to look for a steady source of energy for assured irrigation.

Installation pump sets that work on electricity is not an adequate solution in the majority of the locations as the agricultural fields are far away from the electric grid station. Moreover, the cost of electricity has been increasing gradually and thus increasing the cost of operation of water pumping.

Furthermore, the overall maintenance required and the repair cost of the electric motor used in the water pump is reasonably higher than the solar PV water pumping system. When there was limited knowledge on the benefits of solar PV water pumping system, then for farmers, the diesel pumping system is their first choice as it is portable and economical in terms of maintenance costs.

During that time, the cost of diesel was also cheaper. But it contributed to global warming and environmental pollution by releasing a large proportion of carbon dioxide into the atmosphere. The overall maintenance and repair cost of diesel pumps comparatively higher when compared with solar PV water pumping systems.

Hence, in the current time frame, the solar PV water pumping system is considered the best option for farmers as its pumping cost is cheaper as compared to diesel or electric pump sets.

In addition to this, the contribution of the solar PV water pumping system is less and its maintenance and repair cost is very nominal. It is also highly portable as it can be fitted at any location, whether on the rooftop or the land according to the needs of the farmers because solar energy is innumerable and free of cost in nature.

3.5. SOLAR POWERED MICRO IRRIGATION (SPMI) SYSTEM FOR AGRICULTURE

In broad-spectrum, micro-irrigation can be defined as a pressurized irrigation system that requires gasoline/electrically powered centrifugal pump-set to produce unswervingly pressure for irrigation of crops chiefly for about 16 to 20 hours to produce crop on a day-to-day basis (Figure 3.1).

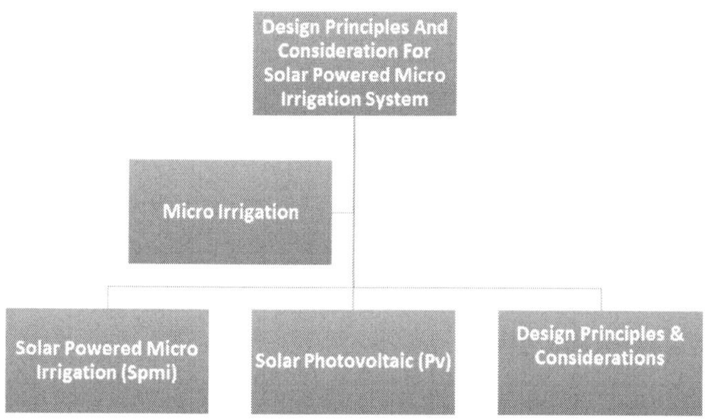

Figure 3.1: Design principle and consideration for solar-powered micro-irrigation system.

This suggests that, while crop irrigation for almost 16 to 20 hours on a daily basis, the recurring cost of fuel or electricity becomes huge resulting in non-viability and non-sustainability of agro irrigation projects. Another major challenge that the farmers faced in the irrigation system is power shortage.

The gap between demand and supply for power is increasing steadily, placing substantial pressure on the government on how to deal with this intricate problem in spite of the allocation of huge funds.

According to UN report, it was reported that there is a deficit of 21% to 85% electricity in major countries and swelling at the rate of 7 to 15% every year. Thus, the majority of attention is given to substitute sustainable and renewable sources of energy, for instance, Solar Power.

Although it is also important to note that solar radiation is freely available, it has its downsides and not so cost-effective to reliable and affordable to everyone.

The concept of solar power has been widely popular over the past few years all over the world in all sectors from spaces station to agro-irrigation due to its infinite availability of free solar radiation without non-judicial and proprietorships.

Thinking about the use of solar power in the micro-irrigation sector is in the primeval stage because of the lack of information and data available on design aspects and remarkable sorts of electro-mechanical machines to pressurized micro-irrigation system.

According to the world data about the solar power data available and potential for solar energy in various countries outline in Table 3.1, it can be concluded that countries positioned across the equator have a wide scope of harvesting maximum solar power @ 350 W/m. while on the other hand, the potential of solar energy decreases while proceeding towards north-south poles @ 3 to 5% per degree of latitude.

Although, there are several factors available on the basis of which solar energy harvesting has relied. Some of these are temperature, clouds, longitude, latitude, relative humidity, and obstructions at specific location.

Thus, it is highly imperative to gather the climatic data from the nearest meteorological station. On the basis of meteorological data, the potential of solar energy harvested can be evaluated for design purposes along with solar modules and allied equipment.

3.5.1. Micro Irrigation

Micro Irrigation refers to a term that is used for portraying the method of irrigation that can be characterized by the following features:

- Water is applied at a low rate;
- Water is applied for a prolonged period;
- Water is applied at recurrent intervals levels;
- Water is directly applied to the root zone of plants;
- Water is applied through a minimal pressure delivery system.

Micro Irrigation System generally conveyed water to the crop by using a piping network and drip lines with drippers spaced at definite intervals along the row of crops is typically driven by diesel engine pump-set and electrical motor pump-set. Each Drippers/emitters supplies a measured, exactly governed by uniform application of water, nutrients, and other required growth substances straight into the root zone of the crops, triggering the growth of the plant and resulting in higher yield.

Table 3.1: Potential of Solar Energy for Various Countries Given for Reference Only

Position	Countries	Solar Power (W/m)
Near Equator	Colombia, North Brazil, Gabon, Congo, Uganda, Kenya, Ecuador, Rwanda, Somalia, Maldives, Indonesia, Tanzania, South Sudan, DRC, Malaysia, Central Africa, Cameroon, Nigeria, Ethiopia, Liberia, Sierra Leon, Ghana, Burkina Faso, Senegal, Venezuela, Costa Rica, Guyana.	300–350
Near Tropic of Cancer	Taiwan, Philippines, Vietnam, Thailand, South China, Burma, India, Bangladesh, Sri Lanka, Pakistan, Nepal, Afghanistan, Iran, Saudi Arabia, Iraq, Israel, Syria, Egypt, Algeria, Tunisia, Jordon, Mexico, Libya, Mauritania, Mali, Niger, Chad, Egypt, UAE, Oman, Myanmar.	250–300
Near Tropic of Capricorn	Botswana, Chile, Madagascar, Brazil, Namibia, Paraguay, Mozambique, and South Africa.	250–300
Above Tropic of Cancer	Japan, North, and South Korea, Turkey, Russia, Ukraine, Germany, Spain, USA, UK, China, France, Italy, Other European countries.	200–250
Below Tropic of Capricorn	Australia, New Zealand, Argentina, Chile	200–250

Source: Table from icid.org.

3.5.2. Solar Powered Micro Irrigation (SPMI)

In Micro Irrigation System that is Solar Power, solar energy (solar PV modules) is mainly used to powered motor pump-set units in contrast to conventional diesel engines or electrical motor pump-set. A typical schematic layout of the solar-powered micro-irrigation (SPMI) system usually embraces a motor pump-set, piping network, solar PV module, drip lines, control head unit, and control panel.

3.5.3. Solar Photovoltaic (SPV)

The PV effect can be stated as photons of light exciting electrons into a higher state of energy, letting them work as charge carriers for an electric current; for example, solar cells transform sunlight energy directly into the flow of electrons. Alexander-Edmond Becquerel (French Physicist) was the first person who observed PV effect in 1839 (Figure 3.2).

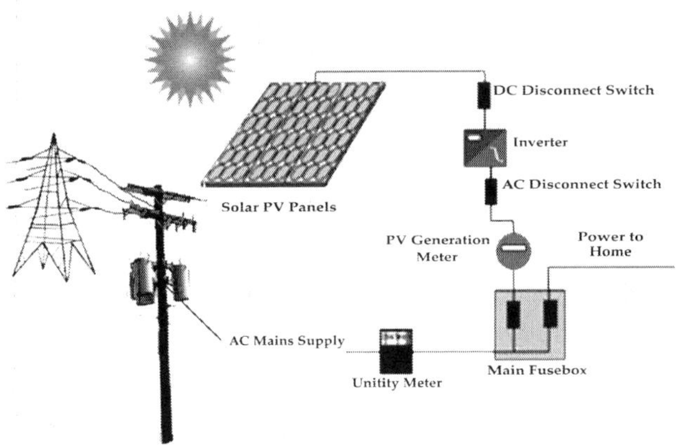

Figure 3.2: Solar PV power plant single line.

Source: Image by Pinterest.

The term PV mainly refers to the unbiased operating mode of a photodiode in which current via the device is especially due to the transduced light energy. Realistically, all types of PV devices are some sort of photodiode. Solar cells generate current direct electricity from sunlight, which can be used to power any equipment or to recharge a battery.

The first practical application of PV was used to give power to spacecraft and orbiting satellites, but today majority of PV modules are used for grid-connected domestic lightings as well as for power generation.

3.5.4. Design Principles and Considerations

Since solar radiation is available in the day time for the generation of power, the entire irrigation activity to grow crops occurs in a day time to efficiently utilize the generated electricity because it is difficult to restore them and used them later.

Therefore, it is always suggested to use solar electricity instantly after its production. Because of this, all the design calculation is based on the variation of solar radiation from morning to evening and from season to season to fulfill the demand of crop water. Although, some factors that is worth considering and integrated while designing micro irrigation system are discussed as follows.

- Day time average solar radiation-on seasonal or on a monthly basis;
- Optimization of ET-crop (overall water demand for crop production) on a daily, alternate, and monthly basis;
- Water assessment and availability to fulfill water demand for crop;
- Management/operation of the system;
- Photovoltaic solar module;
- Analyzing the water retention capacity on the basis of soil quality;
- Master plan of crops;
- Optimization of necessity of motor-pump-set (solar-powered);
- Techno-commercial parameters.

It is worth noticing that proper planning of crops is highly important while selecting a Solar Powered system. Conventional planning of crops is not sufficient and useful and may become costly in overall. Therefore, master planning of crops should be in accord with the demands of crops to a maximum harvest of solar radiation.

This means, there should not be any shortage of solar radiation while the demand of crop water is at its highest stage. Apart from this, there is a need to take some kind of securities such as intensification of soil-based retention capacity by adding organic manure and soil conditioner to minimize the risk of depleting moisture level beyond wilting stage of crops even though there is a gap of irrigation.

Ensuring the proper supply of water is highly important to fulfill the water demand of crops in the daytime when there is an abundance of solar radiation.

Design of micro-irrigation components; for instance, filtration system, piping network, and drip lines, etc., should be satisfactory and corresponding to cope up with the demand of crop at varied solar radiations on a seasonal basis. In another context, the design of micro irrigation system needs to be corresponded with harvested solar power (radiation) to fulfill the crop water demand.

It is also important to note that that average solar energy can be gathered to 200–300 W/min in several countries on the basis of longitude and latitude. It chiefly varies from season to season and mostly affects the haziness and clouds. During the summer season, maximum solar power can be recovered and minimum during cloudiness such as in the rainy season.

Hence, the occurrence of irrigation in the rainy season is relatively low and maximum during the time of summer. In a similar way, in the daytime solar energy is minimum in the morning, while maximum in the noon and again at its minimum during the evening.

Thus, it is right to assume that the micro irrigation system should be designed in such a way that it can help in meeting the water demand of crop in day time corresponding to available solar radiation only.

In the present scenario, there has been noteworthy research and development (R&D) going on to find out the friendly and economical solar-powered system across the world at various levels. Though the initial investment is high, considering the recurring cost and dependency such as the international price of crude oil and currency instability, government policies, the SPMI system is much more economical, sustainable, and deliver a promising future for farmers all over the world to address the power issues, become self-sufficient and independent in food and power.

3.6. PORTABLE SOLAR PHOTOVOLTAIC (SPV) POWERED PUMPING SYSTEM IN VEGETABLE CULTIVATION

The demand for energy has been increasing gradually in every part of the national developments due to the significant growth and expansion in several sectors such as entertainment, agriculture, transportation, irrigation, housing, health, communication, education, industry, and city modernization, etc. (Figure 3.3).

Figure 3.3: Portable solar photovoltaic powered pumping system.

Source: Image by Wikipedia.

To fulfill the increased demand in some countries, particularly in the developing ones, the use of coal-based power plants has been drastically increasing for power generation that poses a severe threat to the environment and is also contributing heavily to global warming by releasing a substantial amount of greenhouse gases and hazardous emissions to the atmosphere.

One only available substitute in this context is to provide backup to the existing power sector with non-conventional energy sources. Among the non-conventional energy sources, solar energy is pondered as the most feasible and attractive one as it is available in copious quality and free to use, mostly available in tropical areas.

In addition, electricity from solar photovoltaic (SPV) systems is gaining a further reputation in the present time frame because of the noteworthy plunge in the cost of SPV modules through recent advancements in this area. The major focus areas for policymakers, planners, and researchers are also now diverted to the applications of SPV system for pumping of water in irrigation sector because of consistent rise in the water demands in agricultural sector and availability of water has become more crucial than ever before.

In developing countries; for instance, in India, diesel and electrically powered water pumping systems are used mostly for irrigation systems. It demands significant energy that is limited in countries like India. Developing a grid system is highly expensive as rural villages are normally located too far away from existing grid lines. Even if there is the availability of fuel within the country, transportation of that fuel to rural and remote areas is quite challenging.

There is no proper infrastructure or road facility to provide much needed supply of fuel in remote areas. Therefore, the use of renewable energy is highly important for water pumping applications in remote areas of many developing countries. While, on the other hand, transportation of renewable energy systems, such as PV pumps, is comparatively easier than the other types because they can be conveyed in pieces and then reassembled at the original site.

Production of PV energy is considered as a core for the production of future energy because it is completely free in its availability, produces no pollution or harmful gases, and is highly reliable. These facts make the PV energy resource more attractive for many applications, particularly in remote and rural areas of developing countries. Solar PV water pumping is widely known for grid-isolated rural locations in places where solar radiation is available in a plethora quantity.

Solar PV water pumping systems can supply adequate water suitable for irrigation activities without the need for any kind of fuel or intensive maintenance as required by electrical and diesel pump sets. They can be installed easily and is durable and highly-reliable to operate. It is also modular, comes with the latest design that makes it easy for future expansion. They can be installed at the site of use, avoiding the spread of infrastructures and long pipelines.

3.6.1. Review of Some Developments in SPV Water Pumping System

The methods for energy security via SPV power system and enhanced water use efficiency (WUE) measures through the use of micro-sprinkler irrigation system in contrast to the traditional flood irrigation.

Thus, it is one of the widely discussed topic among the scientists, researchers, as well as agriculturists not only for ensuring sufficient supply of water to the crops, but also protecting the nonrenewable resources and the environment against noise pollution and release of greenhouse gases by the use of rising diesel pump sets.

Several attempts have been made by some researchers in inspecting the viability of SPV water pumping system for irrigation purposes and domestic drinking water.

Hamidat et al. examined the application of small-scale irrigation with PV water pumping system in Sahara regions. The authors designed a mathematical program to measure the progress of PV arrays under Saharan climatic conditions.

In their research, it was revealed that it is viable to make use of a PV water pumping system for low heads for small scale-irrigation of crops in Algerian Sahara regions. Thus, the PV water pumping system can easily accommodate day to day need of small-scale irrigation with an area smaller than 2 hectares.

It was also concluded that the photovoltaic water pumping system (PVWPS) can improve the overall living conditions of resultant farmers with the advancement of local farming, and thus the migration of the rural workforce will be at its minimum level.

Meah et al. also assessed solar photovoltaic pumping system (SPVPS) for remote locations in rural western USA. They concluded that solar PVWPS (SPVWPS) is an environmentally friendly and at the same time cost-effective way to pump water in remote areas, where 24*7 hours of electrical service is not possible and maintenance is also a serious issue.

From their survey, it was specified that a total of 88 SPVWPS were fitted in all 23 states of the USA, of which 75 systems are still in operation.

They have observed that drought-affected areas such as Montana, Wyoming, Idaho, Oregon, Washington, and parts of Texas can use SPVWPS to enhance their water supply to livestock in remote locations. They were influenced that successful demonstration of these systems is boosting the confidence of other ranchers to use this relatively new technology as another viable water supply option.

They concluded that SPVWPS outperforms other available systems in terms of productivity, reliability, and cost-effectiveness, and the system was able to curtail the carbon dioxide emission substantially over its 25-year life span.

Meah et al. also explored several opportunities of SPV water pumping and challenges in the USA. According to their views, it was established that there is some space for improvements to lessen the initial capital investment requirement and to reduce the maintenance cost and operation services using local-level operation and maintenance.

The authors confirmed that with the help of local resources, such as skills, raw materials, and other valued equipment, the SPVWPS could be economically viable in developing countries and competitive with the conventional diesel generator water pumping systems. It was stated that the SPVWPS should be well-suited with the local practices and culture to accommodate the local needs and wants that can be achieved by using local resources.

3.7. FUTURE TRENDS IN SOLAR POWERED IRRIGATION SYSTEM (FIGURE 3.4)

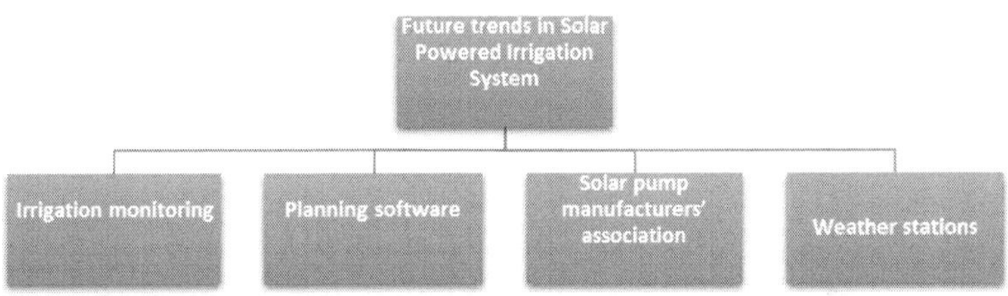

Figure 3.4: Future trends in solar-powered irrigation systems.

3.7.1. Planning Software

It is important to note that in the present day, there are well-designed, tailor-made, and easy-to-use software is available for solar pumping systems and numerous irrigation technologies. Although, there is a need for the integration of technology for irrigation and solar pumps and is expected to be advanced in the near future. A financing option and continuous hard work to assist the formation and execution of these technologies-i.e., without company bias-is promptly required.

3.7.2. Irrigation Monitoring

It is worth noticing that the amount of irrigation water on the field needs to be regulated more frequently and regularly. As the amount of water availability gradually declines, this process will play an even more important role.

For this purpose, a differentiation should be made between monitoring water applied, which can be easily governed in pressurized systems with water transpired and flow meters, that can be evaluated through remote sensing technologies measuring biomass production. Soon, after a period of time, modern technologies will become increasingly common-for instance, supervising irrigated fields with thermal imaging cameras and drones.

Thermal and satellite imagery is already used to compute the demand of irrigation water (and to measure actual supply) for defined areas (e.g., irrigation schemes, fields, watersheds) and to charge farmers correspondingly. An example of a global database is the FAO Water Productivity Open-access Portal (WaPOR) at 30 m, 100 m, and 200 m resolution for certain countries.

3.7.3. Solar Pump Manufacturers' Association

It can be forecast that solar pump companies will set up their platform-for example, a solar pump manufacturers' association. It will help in determining standards for the equipment and make the assessment of information and data possible. In case they will be successful, smaller companies will join the hands to get the advantage from the innovation potential and the data-sharing of such an association.

Many distinctive industries have shifted in this direction, and their stakeholders have profited from the same. One perfect example of this is Regenwassernutzung (fbr) and Fachvereinigung Betriebs, the Association for Rainwater Harvesting and Industrial Water Use in Germany, established in 1995.

It is worth noticing that Rainwater harvesting equipment for industrial and housing buildings was not standardized in the 1980s, and systems were not compatible. Bringing planners, manufacturers, and users join together in an association assists in offering much needed fuel to the sector, to control and standardize the quality of equipment, thus develop it further.

3.7.4. Weather Stations

Weather stations are slowly gaining significance for an optimized irrigation schedule and regime. These stations can be industrialized to a greater extent to become service centers for agricultural crop production as well. It would be achievable only if their databases are advanced to provide not only weather data but data related to soil, such as soil moisture, for the most critical crops of the region.

Forecasting for the upcoming few years could be easily obtainable, so farmers should need to cognizant about how much irrigation water to be transported for each field and crop. It ascertains the requirement for a closely linked network of extended weather stations.

3.8. CONCLUSION

In the end, it is concluded that solar energy in the agriculture sector plays a very important role, especially for developing countries that rely heavily on the power grid for their energy sources. As electric supply is also limited in these countries, particularly in villages, the role of solar energy becomes even more important. One cannot deny the fact that solar energy systems are cost-effective as it does not run on electricity, so it can help in saving down the cost on agriculture produce.

However, the initial cost for setup of a solar system is high, but over a period of time, its overall cost will fall considerably because of no electricity bill. Therefore, this is believed to be the best available option till now if one is looking to implement solar technology in their agriculture sector.

In addition, it should be the duty of the government to provide loan assistance, incentives, and subsidies to poor farmers for helping them to implement a solar system in their farmland. This will ultimately help the government as it no longer needs to provide extra electricity supply to these farmers for several purposes.

REFERENCES

1. Deveci, O., Onkol, M., Ozgur, U. H., & Ozturk, Z., (2015). *Design and Development of Low-Cost Solar Power Drip Irrigation System Using System Monitoring Language*. [Online] ResearchGate. Available at: https://www.researchgate.net/publication/276426978_Design_and_Development_of_a_Low-Cost_Solar_Powered_Drip_Irrigation_System_Using_Systems_Modelling_Language (accessed on 5 May 2021).
2. Energypedia.info. (n.d.). *Sustainable Energy for Pumping and Irrigation-Energypedia.Info*. [Online] Available at: https://energypedia.info/wiki/Sustainable_Energy_for_Pumping_and_Irrigation (accessed on 5 May 2021).
3. Fao.org. (2020). *Solar-Powered Irrigation Systems: A Clean-Energy, Low-Emission Option for Irrigation Development and Modernization*. [Online] Available at: http://www.fao.org/3/bt437e/bt437e.pdf (accessed on 5 May 2021).
4. Frometa, R., (2019). *Seven Benefits of Having Solar Power Irrigation-Vents Magazine*. [Online] Vents Magazine. Available at: https://ventsmagazine.com/2019/09/12/7-benefits-of-having-solar-power-irrigation/ (accessed on 5 May 2021).
5. Goyal, M. R., & Ghosal, M. K., (2020). *Potential Use of Solar Energy and Emerging Technologies in Micro Irrigation*. [Online] Appleacademicpress.com. Available at: http://www.appleacademicpress.com/potential-use-of-solar-energy-and-emerging-technologies-in-micro-irrigation-/9781771883641 (accessed on 5 May 2021).
6. Hartung, H., & Pluschke, L., (2018). *The Benefits and Risks of Solar-Powered Irrigation: A Global Overview*. [eBook] Available at: http://www.fao.org/3/i9047en/I9047EN.pdf (accessed on 5 May 2021).
7. Irena.org. (2016). *Solar Pumping for Irrigation: Improving Livelihoods and Sustainability*. [Online] Available at: https://www.irena.org/-/media/Files/IRENA/Agency/Publication/2016/IRENA_Solar_Pumping_for_Irrigation_2016.pdf (accessed on 5 May 2021).
8. Ososanya, D. T., (2015). *Design and Implementation of a Solar-Powered Smart Irrigation System*. [eBook] Available at: https://www.asee.org/public/conferences/56/papers/13224/download (accessed on 5 May 2021).

9. Rao, K., (2020). *Utilization of Solar Power for Operating Micro Irrigation Systems*. [Online] ResearchGate. Available at: https://www.researchgate.net/publication/331545897_Utilization_of_Solar_Power_for_Operating_Micro_Irrigation_Systems (accessed on 5 May 2021).
10. Roblin, S., (2016). *Solar-Powered Irrigation: A Solution to Water Management in Agriculture?-Renewable Energy Focus*. [Online] Renewableenergyfocus.com. Available at: http://www.renewableenergyfocus.com/view/44586/solar-powered-irrigation-a-solution-to-water-management-in-agriculture/ (accessed on 5 May 2021).
11. *Solar-Based Micro Irrigation System*. (2019). [eBook] Centre for Technology Alternatives for Rural Areas (CTARA) Indian Institute of Technology Bombay. Available at: http://www.ctara.iitb.ac.in/en/system/files/A%20case%20study%20solar%20based%20micro%20irrigation%20system%20.pdf (accessed on 5 May 2021).
12. Sumalatha, A., & Eswar, K. G., (2020). *Energy Efficient Solar Powered Automatic Irrigation System*. [Online] Ijemr.net. Available at: http://www.ijemr.net/DOC/EnergyEfficientSolarPoweredAutomaticIrrigationSystem.PDF (accessed on 5 May 2021).
13. Urja Global Limited, (2019). *Major Benefits of Solar Water Irrigation*. [Online] Available at: http://urjaglobal.in/major-benefits-of-solar-water-irrigation.php (accessed on 5 May 2021).
14. Water Dynamics, (2017). *The Advantages of Solar Water Pumps for Ground Water-Water Dynamics*. [Online] Available at: https://waterdynamics.com.au/advantages-solar-water-pumps-ground-water/ (accessed on 5 May 2021).
15. Whitesharkenergy.in. (2018). *Solar-Powered Drip Irrigation Technique for Agriculture | White shark-Energy, Efficiency, Environment*. [Online] Available at: https://whitesharkenergy.in/blog/solar-powered-drip-irrigation-technique-for-agriculture/ (accessed on 5 May 2021).

Chapter 4

Benefits and Risks of Solar Powered Irrigation

CONTENTS

The chapter "benefits and the risks of the solar-powered irrigation" explain the increasing interest in the solar-powered irrigation system. This chapter also explains the working of the solar irrigation system and where is solar irrigation takes place. This chapter also emphasizes the various components of the solar-powered irrigation system such as motor pump, pump controller, and solar panel. This chapter mentions the various kinds of the motor pumps, such as surface pumps and submersible pumps.

This chapter explains the basic significance of the automatic solar submersible pump control for irrigation along with the components of the automatic solar submersible pump control such as the submersible pump, PV panels, charge controller, battery, inverter, and inverter. This chapter also explains the working of the solar-powered auto irrigation system.

This chapter also addresses the approximated cost to install and run a solar-powered irrigation system. This chapter also provides highlights on the factors that are needed to consider when purchasing a solar power irrigation system such as quality, durability, price, efficiency, warranty, and warranty of the equipment.

This chapter also explains the solar-powered irrigation system toolbox along with its operating principle, components of the solar-powered irrigation toolbox, and the configuration of the solar-powered irrigation system. This chapter also describes the advantages as well as the disadvantages of solar power irrigation systems.

4.1. INTRODUCTION

All across the globe, agriculture is the single largest employer. Agriculture sustains the livelihoods of about 40% of the population all across the world, out of those who continue to live in poverty (United Nations, 2015). Irrigation is between the measures that can enhance yields, decrease the vulnerability to change rainfall patterns and permit several kinds of cropping practices (FAO, 2011).

In this way, irrigation is every so often seen as the engine that provides assistance to make sure the security of food, produces incomes, offers jobs, and drives rural development. Energy is the main input for irrigation services.

As the financial inputs for the solar-powered irrigation systems (SPIS) are decreasing, and subsidy schemes for the solar-powered input systems are being rolled out, the solar techniques are becoming an important option for both large-scale farmers as well as small-scale farmers.

Solar power irrigation system offers reliable as well as cost-effective energy, potentially lesser cost of energy for the process of irrigation. To the rural regions, where diesel fuel is very costly or where reliable access to the electricity grid is absent, they can offer a comparatively flexible and climate-friendly substitute source of energy. The solar-powered irrigation system can be employed in large-scale irrigation systems as well as for decentralized, small-scale irrigation.

In addition to this, several nations are encouraging the solar-powered irrigation system in the framework of national action plans to the sudden changes in the climate as a way, to decrease the emission from agriculture. The operation of the solar pumps does not generate any kind of emission of green-house gas (GHG).

Life cycle assessments of the solar power irrigation systems is responsible for the emission in a cradle-to-grave scenario, which shows a potential decrement in the emission of the GHG per unit of energy which is being used for the water pumping (CO_2-eq/kWh) of 95% to 97% when it is compared with the pumps that operated with the grid electricity (global average energy mix) and 97% to 98%, when they are compared with the diesel pumps (GIZ, 2016).

None the less, it is very important to note that the solar-powered irrigation system, if not sufficiently regulated and managed, it cannot bear the risk of supporting unsustainable water use. Once the installation process of the systems is done, then there is no cost per unit of power and, in this way, no financial profit for farmers to save on fuel or electricity for water pumping.

This can take into the direction towards the wasteful water use, over-abstraction of groundwater, and low field application efficacy. In some instances, farmers sell water to their neighbors at a profit, increment in the overall water withdrawals.

Identifying the risks that are related to the water and coping up with them from the very starting, more particularly in the financing as well as design stages. This will be very crucial or important to make sure the sustainable application of the solar-powered irrigation system technology.

In the context of the quick expansion or rapid development of the solar-powered irrigation system, there is a chance to not simply introduce a clean, climate-smart, as well as innovative energy technology, but to think in a strategic way regarding how this technology can be employed to promote more sustainable use of groundwater resources, to produce more and more inclusive finance as well as management structure and to prosper more combined thinking about the solutions around the water-energy-food nexus.

4.2. INCREASING INTEREST IN SOLAR POWERED IRRIGATION SYSTEM

There is an increment in the interest in solar power irrigation systems all across the globe, noticeable in more and more frequent requests from the agricultural institutions in the emerging nations for installation, finance, and training.

In the year 2015, the Food and Agriculture Organization of the United Nations (FAO) and the Deutsche Gesellschaft für Internationale Zusammenarbeit (GIZ) GmbH have hosted an exploratory workshop to better understand the potential of the solar-powered irrigation system for emerging nations.

Representatives from near about 19 nations shared their experience with the solar pumping technologies, from large scale to small scale, from tropical climate zones to arid climate zones, for vegetable gardens, orchards, and livestock watering, with the help of the surface and groundwater.

It became very clear that there was a requirement to synthesize these kinds of experiences and to respond to the problems, questions, and needs that were arose through the course of the workshop. What is the actual cost of the solar power irrigation system, and what are the advantages of the solar power irrigation system as compared to the other technologies?

What kinds of policies, rules, and regulations are required to handle or cope up with the risks and realize the potential of the solar power irrigation system? What are viable business models for the solar power irrigation system? How can smallholders get the advantages from the technology of solar power irrigation system? How can the risk of groundwater depletion be addressed efficiently?

How can solar power irrigation systems provide assistance to empower women and encourage gender equity? What kinds of capacity development programs are required to support the farmers, extension workers, local private sectors, and others? What are the chances for knowledge exchange and technology transfer? (Figure 4.1).

Figure 4.1: Solar panel energy panels installed in the field.

Source: Image by Pixabay.

FAO of the United Nations and the Deutsche GIZ decided to follow up on these kinds of queries with a global project that is part of the "Powering Agriculture: An Energy Grand Challenge for Development" (PAEGC) initiative. The main objective of the project is to:

- Learn from good practices all across the world;
- Foster policy dialog across sectors on topics, for example, finance, green jobs, and groundwater management;
- Provide assistance to the responsible use of solar power irrigation system with the help of the training and enhanced advisory services.

4.3. SOLAR POWERED IRRIGATION: A SOLUTION TO WATER MANAGEMENT IN AGRICULTURE

Farmer has always played an important role in society, and the reason behind this is that farmers provide food to each and every person all across the globe. Nevertheless, one may forget that, not only the farmers provide food, but in addition to this, the farmers also provide energy, which, in the present interval of time, is of prime importance, more particularly with the respect to renewable energies.

Certainly, farmers can produce energy with the help of the wind, or with the help of the sun, or with the help of the biomass and they can use it for their farm, or, if they have a surplus, resell it to the organizations or firms.

Solar energy can be one of the easiest techniques or methods for farmers to produce energy. Certainly, farmers generally have various kinds of large buildings whose roofs are directly under the sun or there is an adequate amount of solar radiation is present, without being hampered by the shadows of trees, turning them in the form of an ideal place to install a photovoltaic (PV) system.

In this way, the utilization of solar energy in agriculture is becoming more and more common and the energy generated with the help of this renewable source can be used either on the farm or in the local power grid as well, providing the farmer with an additional incentive or additional income.

One of the fields in agriculture that provides advantages the most with the help of solar energy is irrigation, more particularly in dry or parched areas. The main reason behind this is that with the help of the solar radiations or sun for the purpose of irrigation shows a virtuous circle: when the sun shines, it feed the irrigation system, well, it has been known that crops require more amount of water as compared to the amount of the heat. In this way, a huge amount of energy is available when it is required.

4.3.1. How Does a Solar Irrigation System Work?

To transport the water, the pumps are used which are equipped with solar cells. The solar energy absorbed with the help of the cells is then transformed in the form of electrical energy or electric current with the help of a generator which then feeds an electric motor that is used to drive the motor.

Most of the conventional pump systems primarily work on a diesel engine or with the local power grid. Nevertheless, the two modes of operations show disadvantages when they are compared to the solar pumps.

In several numbers of various rural regions, more particularly in developing nations, access to the electricity grid is not always guaranteed. In this case, farmers cannot depend on the conventional irrigation system (Figure 4.2).

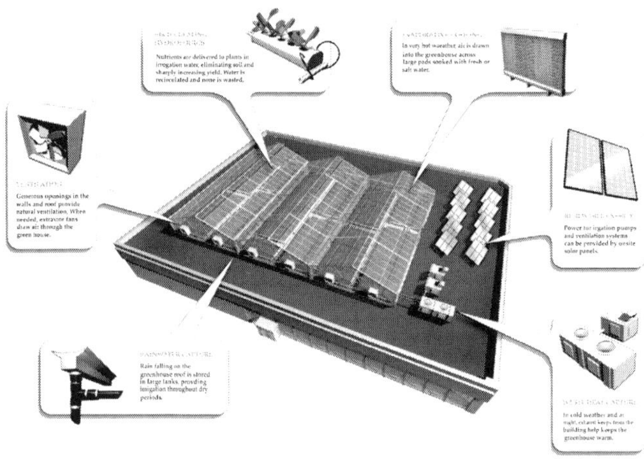

Figure 4.2: An illustration of bright farm system design.

Source: Image by Wikimedia commons.

In this way with the help of an independent and alternative system for providing energy can be a solution for the farmers to safeguard a safe power source and for the public grid to avoid saturation.

Diesel pumps are quite more effective when they are compared with the pumps that work on alternating current. And the reason behind this is that diesel pumps permit greater flexibility. Nevertheless, one of the main restrictions is that this system depends on the availability of the fuel, added to a greater effect on the environment.

Pumps that are driven with the help of diesel are much cheaper when they are compared with the pumps that are driven with the help of solar energy, but the operating costs are quite high and rely heavily on the price of diesel. To the solar-powered systems, they work the other way round, that is, nevertheless this system is comparatively costly, the source of energy is free, in this way, after the amortization period, there are no longer operating costs (only the maintenance costs must be well thought out).

In this way, the solar pumps turn out to be a viable investment for long-term. According to the various studies, for example, Water for wealth and food security by Ag-Water Solutions Project, have depicted, access to the water for agricultural purposes remains critical in some regions, for example, in dry or parched areas of Africa and Southern Asia. Several numbers of farmers from India and Africa fetch the water directly from the well or the rivers and then irrigate the fields with the help of buckets. If the farmers of those regions could have access to a motorized pump, they would elevate their harvesting by near about 300%.

In this way, in the present interval of time, Research and Development tend to concentrate on producing solar pumps that are cost-effective in parched or dry regions. The company named IBC SOLAR has developed with Siemens, which is a solution to substitute the diesel engine with the help of solar-powered engine.

In this case, the entire irrigation system which is also consisting of the pump can remain as it is; only the diesel engine is substituted with the help of a PV system and the so-called "IBC pump drive controller." A prototype of the system was tested in the year 2015 in a farm in Namibia, and as per the manufacture, these systems are turned out to be quite effective and cost-efficient. The prime benefit lies in the fact that there are low acquisition costs as the existing infrastructure is used.

4.3.2. Where Is Solar Irrigation Happening?

The installation of the solar pumps in dry or parched areas, for example, Africa, India, and South America is also part of several numbers of development projects, having the objecting in elevating the productivity of the local farmers and as an outcome, enhancement in their living conditions.

One of the successful instances of this is the initiative of a Physics teacher in a school in Blankenese (Germany) where students have created two pumping systems that work on solar energy in cooperation with the company named as SET GmbH from Wedel (Figure 4.3).

Figure 4.3: An illustration or arid soil.

Source: Image by Peakpx.

They have installed these solar powered systems in two farms in Nicaragua to pump the underground water. This project could also be accomplished with the help of the UNAN University in León, which deals a lot with the use of solar energy.

Certainly, the project has been running for more than 10 years, and about 30 solar-powered pumps are in operation now in Nicaragua. The employment of the solar pumps permits the latter to generate all across the year, even in the dry season and in this way, to elevate their income and support their position in the local market.

Apart from the regions that are mentioned in the above topics, in addition to this, there is an increment in the interest in the solar irrigation system in Europe. Just a few months ago, a mobile solar drip irrigation system from Austria has reached the stage of production.

The Austrian company Wien Energie carried out this project which follows a dual objective. Out of which, the first one is the decrement in the emission of carbon dioxide (CO_2), owing to the use of solar energy. And the second is,accomplishments of near about 30% of water savings thanks to the drip irrigation techniques versus the conventional sprinkler irrigation.

The basic fundamental of the drip irrigation technique is pretty simple. With the help of the several valves, hoes, and pipes, the water drips slowly and at a frequent period of time to the roots of the plants.

In this way, there is no water waste as water goes directly where it should go, unlike the sprinkler system in which the water evaporates into the air or seeps into the soils where no plant grows. In this manner, the drip irrigation techniques permit to harvest of more crops with the help of less amount of water, turning it into a highly effective irrigation method.

For the Wien Energie solar irrigation system, a mobile solar energy system that is equipped with the help of PV modules (up to 3 kW) is linked to a wheeled pump that can pump from wells or rivers.

The pumps that are driven with the help of solar energy then distribute the water directly to the crops with the help of hoses. After a successful test on a 4.5 hectares organic cornfield in Guntramsdorf, Austria, this system is now ready for production.

In this way, in the nations that are suffering from high temperatures, amid regions, and scarce water resources, the drip irrigation system could contribute to effective water management. This is all the more important as farmers have to face three difficulties, that save water, save money, and save energy.

Mobile solar drip irrigation systems shall turn out to be a perfect solution to cope up with these kinds of difficulties. Nevertheless, these kinds of systems are still quite costly and complexed to settle, several numbers of various research and development (R&D) projects are working on the democratization of the use of the solar power in the field of agriculture, which, in the future (and in present interval of time as well), could play an important role in the management of the food and energy crisis.

4.4. COMPONENTS OF THE SOLAR POWERED IRRIGATION SYSTEM

The actual components and the hardware requirement relies on the kind of the irrigation system, for example, surface irrigation where the water is moved across the surface of the agricultural lands, localized irrigation such as spray or drip or trickle system where the water is applied to each and every plant or adjacent to is, and sprinkler irrigation where the water is piped to one central location or more than one central locations with in the field and distributed with the help of an overhead high pressure sprinklers or guns. The main components of a common solar-powered pumping system are consisting of a solar panel array that helps in proving the power to the borewell pump or a surface pump. The size of the entire system is relying on the size of the pump, the amount of water needed per day and the power available from the solar panels (Figure 4.4).

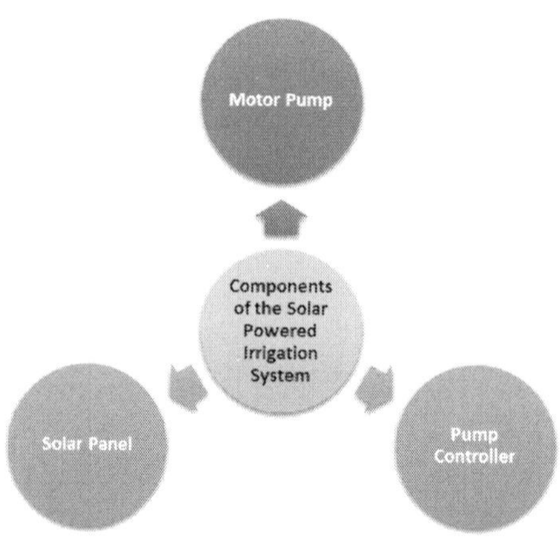

Figure 4.4: Components of the solar-powered irrigation system.

The water is every so often pumped from a borewell or stream into a storage tank or directly into the field. The solar power irrigation system has three main components, that are: Motor pump, pump controller and solar panel.

4.4.1. Motor Pump

The pump is consisting of a motor that runs on the electric power that is produced with the help of the solar panel. According to the type of the motor (that means whether it is alternating current (AC) motor or direct current (DC) motor), the voltage of the solar pump motor can be AC or AC.

DC motors are mostly used for small applications to medium application, for example, garden fountains, landscaping, drinking water for livestock, or small irrigation projects.

Most of the DC pumps run at 24 volts in spite of 12 colts. Nevertheless, smaller systems for 12 volts do exist for light-duty applications. On the other hand, the solar power irrigation systems that are running on 48 volts also exist for much heavier applications. The power of a motor is specified in wither watts or horsepower (HP) (Tables 4.1 and 4.2).

Table 4.1: Typical AC Pump and Controller Specifications

AC Pump HP	Volts	Controller	Solar Panel (watt-peak)	Battery Capacity
1 HP	60	4 kVA	1500 Wp	100 Ah-5 Nos.
1.5 HP	72	5 kVA	1800 Wp	100 Ap-6 Nos.
2 HP	96	6 kVA	2400 Wp	100 Ah-8 Nos.
2.5 HP	96	6 kVA	2400 Wp	120 Ah-8 Nos.

Source: Table from www.solarenergyproducts.co.in.

Table 4.2: DC Pump with Charge Controllers

DC Pump	0.125 HP	0.25 HP	0.5 HP	1 HP	2 HP
Volts	24 V	24 V	24 V	48 V	48 V
No-Load Current	2 A	4 A	8 A	8 A	16 A
Full-load Current	4 A	8 A	15 A	16 A	32 A
Suction	1.83 m (6 Feet)	7.62 m (25 Feet)	16.46 m (54 Feet)	76.2 m (250 Feet)	152.4 m (500 Feet)
Delivery of Flow Rate (liters per hour or minute)	500-lph	8-lph	24-lph	35-lph	70-lph
Panel Capacity	200 Wp	500 Wp	800 Wp	1500 Wp	3000 Wp
Battery Capacity	50 Ah-2 Nos.	150 Ah-2 Nos.	200 Ah-2 Nos.	200 Ah-4 Nos.	200 Ah-8 Nos.
Charge Controller Rating	24 V, 40 A	24 V, 40 A	24 V, 40 A	48 V, 40 A	48 V, 40 A-2 Nos.

Source: Table from www.solar energy products.co.in.

4.4.2. Pump Controller

There are two kinds of pump controllers, the first one is the inverter, and the second is the variable frequency drive (VFD). If an AC solar pump is used, an inverter becomes very significant to alter the DC in the form of AC. The supported power range of a common inverter extends from 015 kW to 55 kW, with the high-powered inverter being used for larger irrigation systems.

The solar panel and the inverter must be sized consequently to accommodate the inrush characteristics of an AC motor. As the pumps that work with the help of AC necessitate high power at the start, the inverter should be able to regulate this extra start-up load.

Every so often, a VFD controller is employed to make sure the motor of the pump gets a sufficient amount of voltage and current. Some typical specifications for several kinds of AC pumps and pump controllers are listed in Table 4.1.

Several numbers of solar DC pumps necessitate a special type of controller if they are to be driven directly with the help of the PV modules (or without the help of the batteries). The controller or linear current booster (maximum power point tracker) allows the pumps to start and run in low light conditions through the course of a rainy day, cloudy day, or early morning and evening. With the help of a battery source, the controller may not be required at all for a DC pump.

A special kind of controller can also be used to transform the 12 volts battery power in the form of a higher voltage to drive the pump at top speed. The DC pumps are available from 12 V to 48 V with their initial torque fluctuating from 12 A to 96 A. A common specification of different DC pumps and charge controllers that are available from a vendor are listed in Table 4.2 as an example.

4.4.3. Solar Panel

The pumps are fitted with motors, which receives energy from the solar PV arrays. The nominal power of a solar module is expressed in peak wattage (Wp). The wattage of the solar panels relies on the necessity and the usage of the motor.

Talking about India, solar PV arrays that are having capacity in the range of near about 200 W to 5 kWp are suggested by the government. A person may refer to the information with respect to the solar PV array and motor pump sets for the solar PV water pumping system (2013–2014) given by the Ministry of New and Renewable Energy, under Jawaharlal Nehru National Solar Mission.

Several numbers of various vendors offer installation services at a cost. But if a person is planning to install by himself, he or she can refer to several numbers of live video clips that are available on the internet.

4.5. TYPES OF MOTOR PUMPS

Solar pumps fall into two major groups, out of which, first is surface pump and second is submersible pump, according to the construction and applications (Figure 4.5).

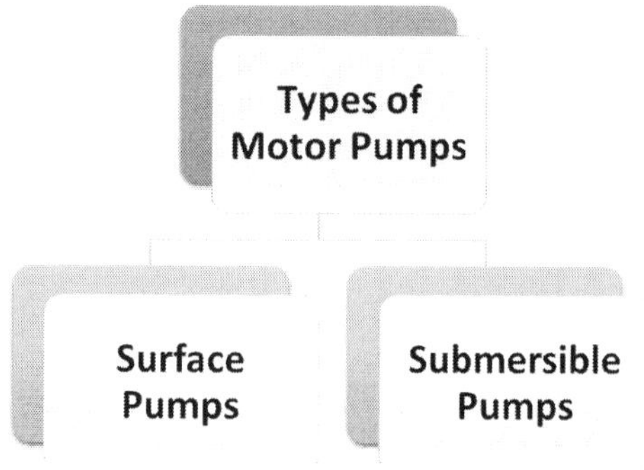

Figure 4.5: Types of motor pumps.

4.5.1. Surface Pump

Generally, the pump is installed above the water level which necessitates a suction lift. A surface pump is used to get discharged from an open well, pond, or river where high flow and small head (height to which the water is required to be lifted) are needed.

There are several numbers of various kinds of pumps that are available, and DC surface pumps with lift up to 21 m (70 feet) are easily available in India.

4.5.2. Submersible Pump

The submersible pump is a hermetically sealed motor. The entire assembly of the submersible pump is submerged inside the water to pump it up. The submersible pump is generally employed for pumping the water from the small-bore wells and is designed to fit inside the drilled bore.

The well may be anywhere from a few meters to near about 550 meters deep. But the deeper the well, the more powerful and expensive would be the system. The 0.25 HP to 3 HP pumps is easily available in the market. DC submersible pumps with a total up to near about 92 meters are quite common in India.

4.6. AUTOMATIC SOLAR SUBMERSIBLE PUMP CONTROL FOR IRRIGATION

These systems work under sunlight. When the sun shines, the water pumping process is a sensible way of solar electric power utilization all across the course of summer, as the requirement of the water is the highest. These kinds of pumps provide a dependable and consistent source for the plantation.

To any kind of pumping system which is based on the solar power, the potential to drive water is a function of three variable for example, power, flow, and pressure. There are several numbers of components are used in this automatic solar submersible pump control panel for irrigation, out of which the main or important components are mentioned as follows.

4.6.1. The Submersible Pump

This solar system is consisting of a submersible pump, an irrigation pump, a water reservoir, and linked water pumps. In the site trail, the submersible pump is kept in a case that is made up with the help of stainless steel, and this case is placed in a well pit at the junction of the open channel and the natural stream course (Figure 4.6).

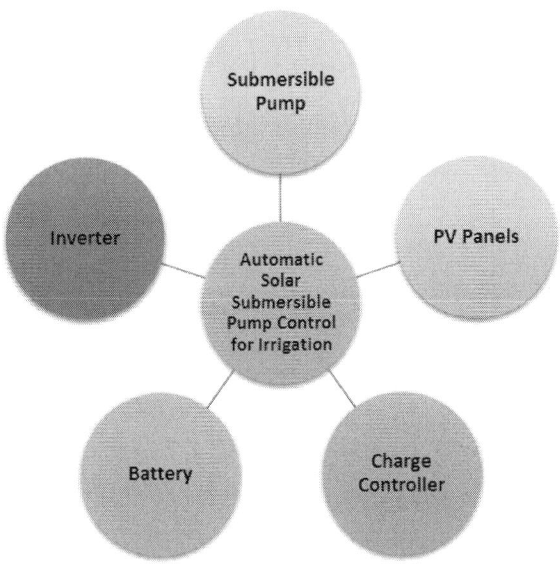

Figure 4.6: Automatic solar submersible pump control for irrigation.

The pump controller pumps the water to the tank at a particular interval of time as set in the control unit. This system is designed with 450 W of power, which is capable of pumping around 2000 liters of water within 60 minutes or in an hour. This power capacity takes into consideration the height difference between the water tank and the solar submersible pump.

4.6.2. The PV Panels

The PV cells rely on the size of the pump. A panel is esteemed in watts of power it can produce. This solar submersible pumping system should be operated with a PV array of capacity in the range of 200 watts to 500 watts peak, and measured under some standard test conditions.

A plenty of modules in series and parallel could be used to get the necessary PV array power output. The output of the PV modules which are used in the PV array standard test conditions should be a main of 74 watts peak.

4.6.3. Charge Controller

A solar charge controller plays a very significant part in any kind of solar power system. The charge controller is used to maintain the sufficient charging voltages of the batteries. The charges controller controls current and voltage from the solar panel and charges the battery, and also stops the charging of the battery from over and undercharging settings.

4.6.4. Battery

The battery is an electric device. The battery is used to store the electric energy which is generated with the help of the solar panel and supplied to the corresponding loads. The number of batteries that are needed relies on the requirement of the load.

4.6.5. Inverter

The prime function of the inverter is that it transforms the voltage of the battery in the form of AC voltage to activate then loads. In this way, it provides assistance to run several numbers of electronic devices, home appliances and computers. In the present interval of time, there are several numbers of various kinds of inverters that are available in the market.

The properties of a conventional inverter are consisting of a high switching frequency, high conversion frequency and less harmonic content, etc.

An instance project for the above automatic water pump controller plant irrigation system is the Solar Powered Auto Irrigation System. The description of this project is mentioned below.

4.6.6. Solar Powered Auto Irrigation System

The prime objective of this project is to create an irrigation system in the field of agriculture with the help of solar energy, and it has several numbers of benefits as well.

The needed components are 8051 series Microcontroller, 12V DC mini submersible pump, Op-Amp, LCD, solar panel, MOSFET, relay, motor, voltage regulator, diodes, capacitors, resistors, LED, crystal, and transistors.

The power supply is made up with the help of a step-down transformer, voltage regulator. In which the step-down transformer provides assistance to step down the voltage to volts AC, and a bridge rectifier which helps in transforming the AC in the form of DC, then a voltage regulator controls the voltage to 5 V which is then employed for the microcontroller operation.

Talking about the solar-powered auto irrigation system project, solar energy is used to activate the irrigation pump. In addition to this, the solar-powered auto irrigation system project is consisting of several sensor parts, which are assembled with the help of an op-amp IC (operational amplifier IC). Op-amps are designed as a comparator. Two coppers wires are injected into the soil to sense the situation of the soil, whether the soil is wet or dry.

A microcontroller in this project is employed to regulate the entire system with the help of noticing the sensors. When the sensors sense the condition of the soil as dry, then with the help of a comparator, the commands are being sent to the microcontroller, and also it sends an instruction to the relay-driver IC then, it reminds the motor to the pump water to the crops.

In this setting, the comparator plays a role of an interface between the sensing arrangement and the microcontroller the status of the soil, and the water pump is displayed on the LCD (which is short form for liquid crystal display), which is interfaced to the microcontroller. In a similar manner, when the sensor senses the condition of the soil as wet, then the microcontroller sends the instruction to the relay to switch off the motor.

4.7. HOW MUCH IT COSTS TO RUN A SOLAR POWERED IRRIGATION SYSTEM?

The cost relies on the factors that are mentioned below:

- Price and number of KWP solar panels needed to cater to the energy requirement;
- Depth of the water source;

- Water discharge requirement or the type of soil;
- Type of motor pump needed (surface or submersible).

The cost of PV panels has intensely reduced through the course of the last 10 years, and it is anticipated to have a more considerable drop in prices in the year 2025.

Apart from this, the other components, for example, pumps and configurations have also been reduced.

Cost approximation for Solar Powered Irrigation System in the period of mid-2017 for Small-scale complete system in Kenya with a submersible pump, 300 W panels on secured 3 m high stand with a controller, filters, and 1-acre drip irrigation, with planning, installation, and guarantees for the pump, panels, and drip lines starts from 2,400 US dollars.

4.8. WHAT TO CONSIDER WHEN BUYING SOLAR PANELS FOR THE SPIS?

Through the course of opting the quality solar panels for the irrigation system, a person or an individual should well thought out about the following criteria that are mentioned in subsections.

4.8.1. Quality

- ISO certified panels indicate quality;
- The International Organization for Standardization (ISO) 9000 series assures the standards of quality for the manufacturing industry.

4.8.2. Durability

- Panels should fulfill the International Electrotechnical Commission's (IEC) 61215 reliability standard to be considered durable;
- Solar panels are engineered to survive in extreme settings such as harsh winters and strong winds.

4.8.3. Price

- The cost of solar panels is equal to the price per watts plus installation and permits;

- It is not guaranteed and essential that the most expensive panels are the best ones;
- The average solar panel cost in 2020 is near about $ 2.99/watt (Figure 4.7).

Figure 4.7: Things to consider when purchasing solar panels for solar-powered irrigation systems.

4.8.4. Efficiency

1. Solar panels can transform the solar radiation in the form of usable electric energy;
2. The most effective solar panels of 2020 are the Sun-Power X-Series Solar Panels with a record-breaking of 22.80%;
3. Higher efficiency rating is better, and the reason behind this is that it means that more electricity is transformed with the help of solar radiation to generate up to 75% more energy per square foot over the period of first 25 years;
4. The top solar panel manufacturers by efficiency are Sun Power, LG, REC Group, Panasonic, and Solar tech Universal.

4.8.5. Warranty

As solar panels is a long-term and costly investment, it is very necessary to know that Solar panels have two warranties, as mentioned below:

- Performance Warranty-Solar panels degrade over the interval of time just like any other product; and
- The standard performance warranty of the industry is around 25 years.

Performance warranty makes sure that the product will perform and generate electric energy at a certain level of efficiency after a particular number of years.

4.8.6. Equipment Warranty

- It covers any physical faults or issues that will arise from the manufacturing errors for example glass imperfections, damaged cell connections, loose junction boxes, and so on;
- A standard product warranty is near about 10 years to 12 years;
- Some products offer 25 years of warranty such as Sun Power, Solaria, LG, Panasonic, and Silfab;
- It is better to consult or work with a solar energy consultant before purchasing and installing the solar panels, and the reason behind this is that they are professionals and they know much more about this industry.

4.9. SOLAR POWERED IRRIGATION SYSTEM TOOLBOX

With the help of solar energy for irrigation makes a lot of sense. Out of which, the first is, irrigation is every so often in rural regions with poor access to reliable electric energy or supplies of fossil fuel. The second reason is, solar radiation or sunlight is a never-ending resource, more particularly in the areas where rainwater scarcity makes irrigation important in the context of the food industry and international trade (Figure 4.8).

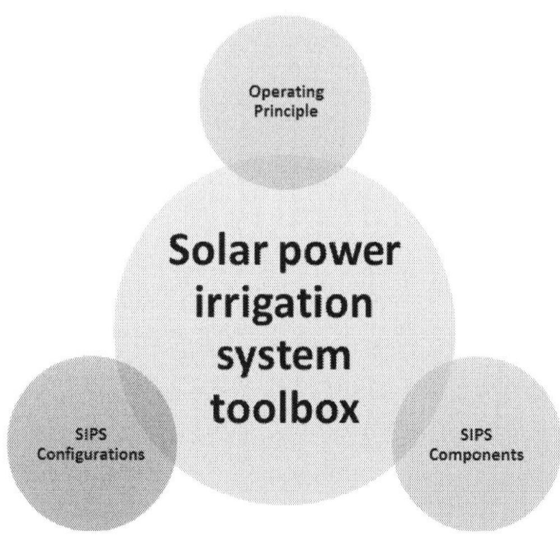

Figure 4.8: Solar power irrigation system toolbox.

In the end, SPIS passively self-regulates, due to the volume of the water pumped elevates on a clear hot day when plants necessitate more amount of water, and vice-versa. It is very necessary in the order note that SPIS is more than just a solar pump that are used for irrigation purpose.

Panels, pumps, and irrigation systems are designed on the basis of the availability of the water and the requirements of the local crop water. SPIS is a system where the several numbers of various components, from pump to plant, are combined and harmonized.

4.9.1. Operating Principle

The operating principle of SPIS is simple. A solar generator gives electric energy to an electric motor pump, which transports the water either directly into an irrigation system or to increase the level of water in the reservoir. Basic design criteria for SPIS are consisting of minimum maintenance, maximum reliability, along with resource efficiency.

A specific aspect of SPIS is the fact that generally a battery backup is not needed. This is a benefit, and the reason behind this is, batteries are maintenance-intensive, expensive and necessitate frequent replacement.

4.9.2. Components of Solar Powered Irrigation Systems (SPIS)

The individual components of a SPIS are introduced in the following sections. Table 4.3 shows that, according to site-specific conditions and the potential of the farmer, several kinds of technology options are available. The components along with the alternative technologies can be integrated in an extensive range of ways, but some configurations are better according to the situation in the field.

Table 4.3: Major Alternative Technologies for the Most Important Components of an SPIS

Component	Alternative Technologies		Dependent On:
Solar system	Fixed	Tracking	Costs and maintenance intensity
Pump	Surface	Submersible	Costs and (geo-)hydrology
Reservoir	Reservoir	No reservoir	Costs and irrigation system
Irrigation system	Surface	Drip or sprinkler	Costs and pumping system

Source: Table from energypedia.info.

4.9.3. Solar Power Irrigation System Configurations

The most common solar-powered irrigation system configuration is when a solar generator on a fixed mounting structure offers electricity for a submersible pump that is installed in the borehole. The water is then pumped into the reservoir which is elevated a few meters above the field where it is stored at a constant pressure.

When it is released, it flows into a drip irrigation system, which works on the low pressure, and where the water is being filtered and being mixed with the fertilizer before reaching the plants.

Nevertheless, the installation of the water filter on the exit or output side of the tank can be a cause of the problem, because the pressure losses in the filter can easily reach several meters, and then, at the low tank heights, there is no water flowing any longer. In this way, it is recommended to install the filter on the input side, to keep clean water in the tank.

This configuration would also be possible with the help of a tracking system but would necessitate a higher financial input and more maintenance as compared to a fixed mounting of the solar panels. The reservoir offers stable pressure and water supplies to the drip irrigation system to make water distribution as uniform as possible.

The performance of drip irrigation reduces as the drippers get clogged because of the small particles that are present in the water. Filters prevent this, but only when properly designed for the specific kind of quality of water and irrigation system, and only when is cleaned frequently.

In this way, it is strongly referred to employ drip irrigation only in the combination with the groundwater as the groundwater is much cleaner as compared to the water which comes from the rivers or open reservoirs. In addition to this, it is suggested to have a monitoring system installed between the pump and the reservoir to measure the flow of water and the pressure of the water.

The simplest solar power irrigation system configuration is when a solar generator on a fixed mounting structure offers electricity for a surface pump that is installed at the reservoir or river. The water is then pumped directly into the surface of the irrigation system, for example, with the help of a network of open canals. For this configuration, the pumped water does not pass through an elevated reservoir.

The pressure of the water as well as the flow of the water to the irrigation system corresponds to the actual solar irradiance, which fluctuates over the day, more particularly, with a fixed mounted solar generator. The main benefit of this configuration is the simple installation and comparatively low costs.

Nevertheless, the main limitation of this configuration is that the producer has little command on the distribution of the water in the field over the days because no reservoir controls the flow of the water as well as the pressure of the water. The producer will have to use volumetric valves or separate his field into manageable sub-sections to regulate the watering of the crops as best as possible.

4.10. ADVANTAGES OF SOLAR POWERED IRRIGATION

Most recently, several numbers of farmers are turning towards the use of solar energy for all their irrigation requirements, and it is so with good reason. Farming can be quite an expensive endeavor, more particularly, when done on a large scale (Figure 4.9).

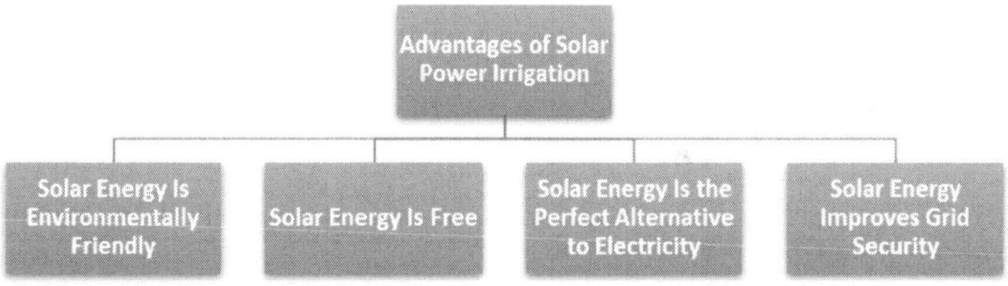

Figure 4.9: Advantages of the solar power irrigation.

In this way, the incorporation of solar irrigation goes a long way in unbalancing some of those over-the-head costs, which can, as an outcome, be channeled to the other more critical sectors of the farm. There are various numbers of reasons that a person is required to have solar irrigation installed on the farm, which is discussed in subsections.

4.10.1. Solar Energy Is Environmentally Friendly

In spite of making sure that the crops are well watered each and every day, solar energy is also known to be environmentally friendly. Not similar to the several energy sources, for example, biogas, and coal, solar power does not emit greenhouse gases. As a person already know, greenhouse gases have wreaked some critical chaos on the ozone layer for the past several numbers of years.

Solar irrigation is also ideal, and the reason behind this is that it does not produce noise as is the case with the generators. In the end, solar energy does not necessitate that much water for maintenance purpose. In simple words, a person will end up saving on the water which can be used to irrigate the crops.

4.10.2. Solar Energy Is Free

Not like electricity, where a person is obligated to make monthly payments, solar energy is free. A person is required to have a solar panel which is installed on the form, and then a person is good to go. In addition to this, make sure the person is investing in the best quality of the solar panels that are installed with the help of a seasoned expert.

4.10.3. Solar Energy Is the Perfect Alternative to Electricity

Farmers who use the electricity to run their irrigation system know exactly how disrupting an impromptu blackout can be. As the generators are very costly to both run as well as maintain, solar energy is every so often seen as the perfect, which is the most efficient substitute. And the reason behind this is that it not only runs on the sunlight, which is free, but the system is also quite straight forward from an operational standpoint.

4.10.4. The Solar Energy Improves Grid Security

The grid will always be safeguarded as long as it has solar energy penetration. This explains why a network with power plants that are spread out will have experience very few blackouts. The latter will also play a very important role in safeguarding the grid in the action of a natural disaster, some of which can take place out of the blue.

4.11. DISADVANTAGES OF SOLAR POWER IRRIGATION (FIGURE 4.10)

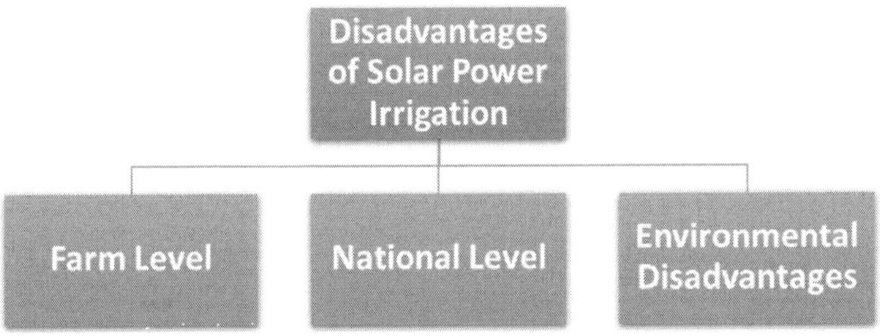

Figure 4.10: Disadvantages of the solar power irrigation.

4.11.1. Farm Level

- Still comparatively high initial investment costs that smallholder farmers, more particularly, cannot afford or cannot tolerate the dangers come along with the investment.
- Finance is not accessible or affordable for all, more particularly for the smallholder and tenant farmers.
- Design requires to be fit-for-purpose and necessitates services (generally a private sector) to advise farmers on best system; nevertheless, these are every so often not in place.
- Optimal operation and maintenance of Solar Power Irrigation System necessitate a certain amount of technical knowledge and skill, in this manner, the farmers are required to be trained, and services (extension services or private service suppliers) need to be available.
- Solar Power Irrigation System are vulnerable to theft, and in this way, they, every so often not covered with the help of insurance as a prerequisite for loan finance.
- Deficiency of trust between the farmers, utilities, service providers and government to try innovative forms of finance and of FITs; banks every so often perceive that Solar Powered Irrigation System has high risk, and the reason behind this is to unfamiliarity with the technology.

4.11.2. National Level

- Existing energy subsidies for fossil fuels and electricity that distort the market; legislation and regulation of energy and agricultural markets may hamper the uptake and up-scaling of solar energy systems. Taxes on imported equipment may distort (and artificially keep up) prices.
- High risk investment, more particularly, if the solar-powered irrigation system roll-out programs do not sufficiently address onsite ownership.
- Deficiency of groundwater management/institutional framework for abstraction.
- Deficiency of codes and standards to guarantee the quality of solar-powered irrigation system.
- Deficiency of systematic training schemes (for example, vocational training); in this way, deficiency of skilled personnel.

4.11.3. Environmental Disadvantages

- Production of photovoltaic panels necessitates some toxins as well as rare minerals; mining and production of these tends to cause the environment by producing the environmentally harmful waste; panels are required to be correctly disposed of to avoid environmental dangers.
- Decentralized systems are very challenging task to control (nevertheless, solar-powered irrigation system, every so often substitute already decentralized systems); illegal pumping may elevate.
- Risk of groundwater over-abstraction, that takes into the direction towards the depletion and degradation of groundwater resources.

4.12. CONCLUSION

In the conclusion of the chapter, the agricultural farm is the backbone of almost every local economic development, more particularly in the emerging nations. And there are three factors in the field of agriculture that are considered as most important, that are soil, water, and sun.

Sun is replaceable by artificial light. Soil is replaceable by hydroponics. But there is no alternative to the water. Water irrigation is very necessary in the case of a deficiency of rain. There are some important factors that are needed to be considered to accomplish or achieve farm, for example, reliability, affordability, and access to irrigation water.

More particularly, the farmers who live in rural regions have no access to a reliable source of water for the process of irrigation. And if there is access, they are not able to afford it. One promising solution to the related issue, considering these aspects, is the solar-powered irrigation system.

Solar power irrigation system (SPIS) is an automatic irrigation system. In a Solar-powered irrigation system, the irrigation pump is driven with the help of the electric energy that is produced with the help of solar radiation or sunlight, which is transformed with the help of solar panels or PV cells.

Installation of the solar-powered irrigation system is an expensive investment, but on the other hand, it is a cost effective-talking in long term as compared to the conventional water pumps that use non-renewable fossils such as diesel and gas that contributes to further harms to the environment.

This technological option becomes very familiar and well-known, and the reason behind this is that, a noticeable reduction in the panel prices and the cooperation or help of governments as well as technical extension services.

The solar-powered irrigation system has a certain importance, not only to the economy of a country but most particularly to the environment.

REFERENCES

1. Adnalandal, (2019). *Major Benefits of Solar Water Irrigation*. [Online] Urja Global Limited. Available at: http://urjaglobal.in/major-benefits-of-solar-water-irrigation.php (accessed on 5 May 2021).
2. Electronics for You, (2014). *Solar-Powered Irrigation Systems | Electronics for You*. [Online] Electronics for You. Available at: https://www.electronicsforu.com/technology-trends/tech-focus/solar-powered-irrigation-systems (accessed on 5 May 2021).
3. Energy Pedia, (2018). *SPIS Toolbox-Solar-Powered Irrigation Systems-Energypedia.Info*. [Online] Energypedia.info. Available at: https://energypedia.info/wiki/SPIS_Toolbox_-_Solar_-_powered_Irrigation_Systems (accessed on 5 May 2021).
4. Green Match, (2020). *Advantages and Disadvantages of Solar Energy (2020) | Green match*. [Online] Greenmatch.co.uk. Available at: https://www.greenmatch.co.uk/blog/2014/08/5-advantages-and-5-disadvantages-of-solar-energy (accessed on 5 May 2021).
5. Hans Hartung, (2018). *The Benefits and Risks of Solar-Powered Irrigation: A Global Overview*. [eBook] the Food and Agriculture Organization of the United Nations and Deutsche Gesellschaft für Internationale Zusammenarbeit. Available at: http://www.fao.org/3/i9047en/I9047EN.pdf (accessed on 5 May 2021).
6. Helen, D., (2019). *What is Solar Irrigation?-Futurepump*. [Online] Futurepump. Available at: https://futurepump.com/what-is-solar-irrigation/ (accessed on 5 May 2021).
7. Nevon Projects, (2019). *Irrigation System running on Solar Power*. [Online] Nevon Projects. Available at: https://nevonprojects.com/irrigation-system-based-on-solar/ (accessed on 5 May 2021).
8. Renewable Resources Coalition, (2016). *Solar Energy Disadvantages: The Top Drawbacks of Solar Power*. [Online] Renewable Resources Coalition. Available at: https://www.renewableresourcescoalition.org/solar-energy-disadvantages/ (accessed on 5 May 2021).
9. RJ Frometa, (2019). *Seven Benefits of Having Solar Power Irrigation-Vents Magazine*. [Online] Vents Magazine. Available at: https://ventsmagazine.com/2019/09/12/7-benefits-of-having-solar-power-irrigation/ (accessed on 5 May 2021).

10. Solar Power Nerd, (2020). *What is a Solar-Powered Irrigation System? | Solar Power Nerd.* [Online] Solar Power Nerd. Available at: https://solarpowernerd.com/what-is-solar-power-irrigation-system/ (accessed on 5 May 2021).
11. Stéphanie, R., (2016). *Solar-Powered Irrigation: A Solution to Water Management in Agriculture? Renewable Energy Focus.* [Online] Renewableenergyfocus.com. Available at: http://www.renewableenergyfocus.com/view/44586/solar-powered-irrigation-a-solution-to-water-management-in-agriculture/ (accessed on 5 May 2021).
12. Tarun, A., (2020). *Solar Powered Automatic Irrigation System-ECE Projects.* [Online] ElProCus-Electronic projects for engineering students. Available at: https://www.elprocus.com/solar-powered-automatic-irrigation-system/ (accessed on 5 May 2021).

Chapter

5

An Overview of Wind Energy

CONTENTS

Wind Energy is one of the cheapest and easiest ways to produce energy. In the chapter, an overview of wind energy, the way how wind energy is produced is discussed. The chapter also discussed several challenges and advantages in the production of wind energy.

Several characteristics of wind energy are also discussed in this chapter. It also shed some light on the different types of wind energy. In addition, it also discussed the history of wind applications. It also stated the importance of the integration of wind and other energy sources. It also highlighted grids, markets, and system integration.

5.1. INTRODUCTION

Wind energy is one of the forms of solar energy. It describes the process by which there is the use of wind in the generation of electricity. Wind turbines help convert the wind's kinetic energy into mechanical power. A generator is useful in the conversion of mechanical power into electricity. The utilization of mechanical power can also be used directly for particular tasks such as pumping water. The phenomenon of wind can occur by the uneven heating of the atmosphere by the sun, variation in the surface of the earth, and earth's rotation. Mountains, water bodies, and vegetation altogether influence wind flow patterns. Wind turbines convert wind energy into electricity by the rotation of propeller-like blades around a rotor. The rotor switches the drive shaft, which turns an electric generator. Three primary factors that affect the quantity of energy a wind turbine can extract from the wind include air density, wind speed, and swept area (Figure 5.1).

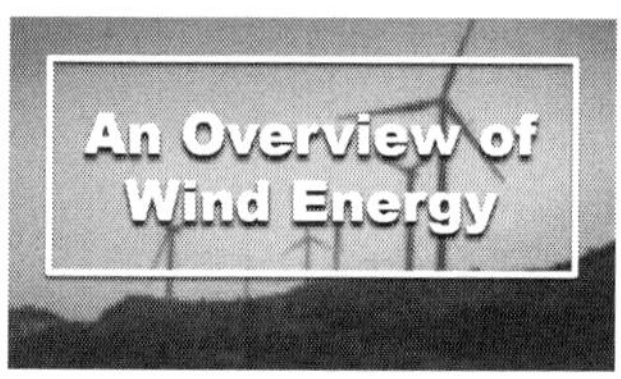

Figure 5.1: An overview of wind energy.

Wind power is growing rapidly and is one of the promising renewable energy technologies. The usage of this technology is rising in the world, and one of the primary reasons is low cost. Globally, the onshore and offshore wind generation capacity has increased by 75 in the last two decades. Many sites of the world contain strong wind speeds, but the best places for wind generation are remote areas usually. Offshore wind power provides tremendous potential.

The first trace of wind turbine emergence is more than a century ago. With the invention of the electric generator in the 1830s, engineers started researching wind energy for electricity production. Wind power generation begins in the United Kingdom and the United States in 1887 and 1888 respectively, but modern wind power's first development is believed to take place in Denmark, wherein 1891, horizontal-axis wind turbines were built and a 22.8-meters wide turbine began to operate in 1897.

Wind produces electricity by the use of kinetic energy created by air in motion. It transforms into electrical energy by using wind energy conversion systems (wind turbines). The working system of these turbines includes wind hitting the turbine's blades and making them rotate and switch the connected turbine. It changes the kinetic energy to rotational energy, by moving a shaft that is generator-connected, and thereby producing electrical energy by electromagnetism.

The amount of power harnessed from the wind depends on the side and length of turbine blades. The result is directly proportional to the rotor's dimensions and the cube of the wind speed. Engineers believe that when the wind speed doubles, the potential of wind power increases by a factor of 8.

Over time, one can see an increase in wind-turbine capacity. In 1895, the rated capacity of typical turbines was 0.05 MW (megawatts) and a 15 meters' rotor diameter. In the present scenario, new wind power projects have a turbine capacity of 2MW onshore and 3 to 5 MW offshore.

Commercial wind turbines have reached the capacity of 8 MW with rotor diameters reaching up to 164 meters. There is an increase in the average capacity of wind turbines from 1.6 MW in 2009 to 2 MW in 2014.

Due to its large number of benefits, wind power is one of the fastest-growing energy technologies in the world. It also presents inherent challenges in some areas of the world, Research and Development (R&D) projects address those hurdles around the globe.

In many regions, wind power is cost-effective while it needs to compete with other energy sources in others. Global R&D efforts are currently working on solutions that will reduce the LCOE (Levelized cost of electricity) of onshore and offshore wind power.

One of the biggest potentials of wind power is that it is a domestic source of energy, hence provides a limitless local resource. Some applicable wind farm locations are remote areas that come with challenges in construction and energy transmission logistics. New technologies coming of age technologies like two-piece blades and modular construction are helping in overcoming such challenges.

An additional wind power benefit is that it is a sustainable energy source. Wind turbine operation does not emit greenhouse gas or CO_2. It helps countries meet their targets of emission reduction and combating climate change. Wind energy is readily available, plentiful, and capturing it does not cause depletion in our valuable natural resources.

The primary environmental benefit of wind power is its ability to wafer the detrimental effects of climate change. According to Global Wind Energy Outlook, by 2030 the wind energy projects will offset 2.5 billion tons of carbon per year.

5.2. WIND ENERGY FOR DEVELOPMENT

In the 1980s, the first commercial wind energy converters came into existence when the total installed wind generation capacity in the world was only 5,000 MW, although the wind energy boom as such did not begin until the mid-1990s. The installed capacity has increased at double-digit rates of every 12 months growth within a short period.

Global installed capacity had progressed to 74,233 MW by the end of the year 2006. Installed globally as of 2011, at present, the industry is enjoying a boom with 239,000 MW. The installed systems are required for generating electricity almost without any exception. At present, Europe is still the largest market where some 48,545 MW (65%) is already been installed and out of this, 22,000 MW is located in Germany (figures from end of year 2006) (Figure 5.2).

Figure 5.2: Wind energy for development.

In the system manufacturers, Germany is also a lord and master. Among the major manufacturers of the world, there are four German companies and the German component industry supplies clutches, gearboxes, and other assemblies to a great number of producers in several other countries.

It is beyond doubt that the wind industry has made remarkable progress even if it remains a matter of dispute whether wind energy would still be competitive without promotional support. In the early 1990s, specific investment costs have fallen to around 900 EUR/kW while in the meantime the cost of systems still averaged almost 1,300 EUR/kW.

By considerable increases in the efficiency of turbines (larger rotor diameter, greater hub height, etc.), the merits of mass production have been further boosted, which have improved the economics of wind energy. Currently, there are turbines with a rated output of up to 6 MW in the market. It is important to consider that the growing market in the wind industry in electricity generation and grid feed-in is noticeable.

- **Potentials:** In many developing as well as emerging countries, the wind energy potential is substantial. There are numerous locations where generating electricity out of wind energy displays an economically viable alternative as compared to conventional fossil energy sources, for example, diesel or coal. Wind turbines are an alternative to conventional power stations in developing and emerging countries.

In many places, wind energy is now considered to be cost-effective as well as non-polluting and reducing dependence on imports of fossil fuels in comparison to fossil-fueled power stations.

5.2.1. Advantages of Wind Energy

- Wind energy involves usage of an indigenous resource in the absence of producing any type of pollution or greenhouse gases;
- It gives rise to the power supply diversification;
- Local resources in terms of capital, labor, and materials are developed by wind energy projects;
- Wind projects can be included in the new approach of independent power production (IPP) and attract new capital;
- Projects of wind give a boost to the cooperation with different donors involving Germany, enhancing local capacities as well as technological knowledge.

5.2.2. Challenges of Wind Energy

- In developing and emerging countries, even good wind resources have not been used to the desirable extent despite the economic and ecological advantages;
- From the point of view of global wind energy companies, the barriers for private investment are also a great challenge besides the difficulties of raising capital and risk covering;
- In the developing as well as emerging countries, the essential reasons for this depend on the lack of knowledge;
- Inadequate information about the energy-sector framework conditions as well as support mechanisms;
- Insufficient information about global markets;
- Lack of legal framework of wind energy (economical and technical conditions for feeding wind-generated electricity into permit procedure, power grids, etc.);
- Availability of qualified staff, especially in the field of service and maintenance is also one of the challenges. In remote locations, installments often break down because of a lack of servicing, spare parts, or trained workforce to administer them, and technicians as well as buyers are often unfamiliar with wind technology.

In fact, windpumps are less maintenance intensive as compared to diesel pumps. In spite of that, many people are not aware of the windpump technology, and hence there is a need to train maintenance staff in the required areas.

- Lack of developed infrastructure required to assist the installation, commissioning as well as maintenance of wind generators. Often technicians and users are unaccustomed to the technology.
- Cost of Investment: The investment cost of purchasing a windpump is usually higher than that of diesel pumps however, the lifetime cost of wind is often less than diesel- or petrol-powered pumps. Generally, buyers of water supplies possess limited funds and are unable to take a long-term view of the technology.
- Fuel-fired power plants have consistent output as compared to wind energy. To allow usage in periods of low or no winds, small-scale wind generators require battery storage. A stable grid is required to act as the storage for grid-connected systems. Wind pumps need storage of water.

- To work over a given range of wind speeds usually 4–12 m/s, wind generators are designed. In simple terms, the technology can only be used in areas with sufficient winds.

The use of wind was to move the sails of the ships in ancient times. In this chapter, it has been discussed that how to generate electricity with the usage of wind energy.

A turbine transforms the wind's kinetic energy into useful mechanical energy. Mechanical form or turn generator turbines are some of the areas where this energy is used and hence provide electricity. Wind energy harnesses by the conversion of the wind kinetic energy to mechanical energy, just like in hydropower systems.

There are two types of wind turbines and that are Horizontal Axis Wind Turbines and Vertical Axis Wind Turbines. Installation of wind turbines in large areas, in other words, wind farms are increasingly emerging today.

5.2.3. Wind Characteristics

The wind has many general characteristics, while others are more specific to the site. A few of the site-specific characteristics involve:

- **Mean Wind Speed:** The annual wind yield is estimated by this however it does not give the distributions.
- **Wind Speed Distribution:** Annual, diurnal, and seasonal characteristics are its three aspects. It is necessary to understand the variations and the spread of wind speed while choosing a site.
- **Turbulence:** In unpredictable patterns, this is the chaotic movement of the wind. Turbulence is the result of continuous changing properties of wind motion that affect the production of energy as well as fatigue on blades.
- **Long-Term Fluctuation:** Unpredictable energy supply is the cause of irregular wind. It is necessary to study the area for a constant wind flux before a wind turbine is set.
- **Distribution of Wind Direction:** This is very important while setting the position of the blades, especially for horizontal axis types as compared to vertical axis wind turbines.
- **Wind Shear:** Shear refers to the change in wind speed, direction, or the height at which the maximum velocity takes place.

5.2.4. Wind Speed Patterns

Generally, wind patterns are important and analyzed with the help of a wind spectrum. A high value of the spectrum of wind transforms the speed of wind at the given time interval. The peaks represent turbulences that take place with time in the case represented on a graph.

5.2.5. Wind Speed Distribution

There are three types of distributions:

- **Diurnal:** This type of wind speed distribution is the result of the difference between temperatures during the day as well as at night.
- **Depressions:** It takes place with an interval of four days along the coastal region.
- **Annual:** Distribution is dependent on latitude.

5.3. TYPES OF WIND ENERGY

Wind occurs when the sun heats the surface of the earth unevenly. To generate electricity, wind energy can be used (Figure 5.3).

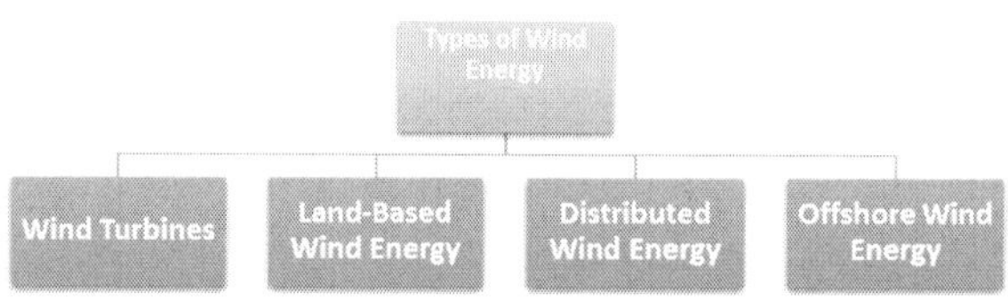

Figure 5.3: Types of wind energy.

5.3.1. Wind Turbines

Wind turbines, such as windmills, are embedded on the top of the tower so that it can catch the most energy. At 100 feet, in other words, 30 meters or more aboveground, they can take advantage of the less turbulent as well as faster wind. Turbines use their propeller-like blades to catch the energy of wind. Usually, a rotor forms when two or three blades are mounted on a shaft. A blade is very similar to an airplane wing. A pocket of low-pressure air forms on the bottom side of the blade when the wind blows. A low-pressure air pocket then pulls the blade towards it. This causes the rotor to turn, which is called lift.

The force of the lift is much stronger as compared to the force of the wind against the front side of the blade. This is known as drag. The combination of drag and lift makes it possible rotor to spin just like a propeller, and the turning shaft spins a generator to make electricity.

5.3.2. Land-Based Wind Energy

Wind turbines can be connected to a utility power grid or even combined with a photovoltaic (PV) (solar cell) system. Standalone applications are also one of its uses. A large number of wind turbines are usually built close together to form a wind plant for utility-scale (megawatt-sized) sources of wind energy, also referred to as a wind farm. Nowadays, a large number of electricity providers use wind plants so that they can supply power to their customers.

In most cases, uses of standalone wind turbines are for communications or pumping water. However, in windy areas, homeowners, farmers, and ranchers often use wind turbines as a way to cut their electric bills.

5.3.3. Distributed Wind Energy

Distributed energy resources also have potentialof small wind systems. Distributed energy resources denote varieties of small as well as modular power-generating technologies combined to upgrade the operation of the electricity delivery system.

Distributed wind energy systems, commonly installed on, but are not limited to residential, agricultural, commercial, industrial, and community sites and can range in size from a 5-kilowatt turbine at a home to multi-megawatt turbines at a manufacturing facility or connected to a local distribution system.

5.3.4. Offshore Wind Energy

In the United States, offshore wind energy is relatively a new industry. The first offshore wind farm of America that is located in Rhode Island, off the coast of Block Island, was powered up in December 2016. Wind Vision Report of the Energy Department shows that offshore wind could be available in all coastal regions nationwide by the year 2050 (Figure 5.4).

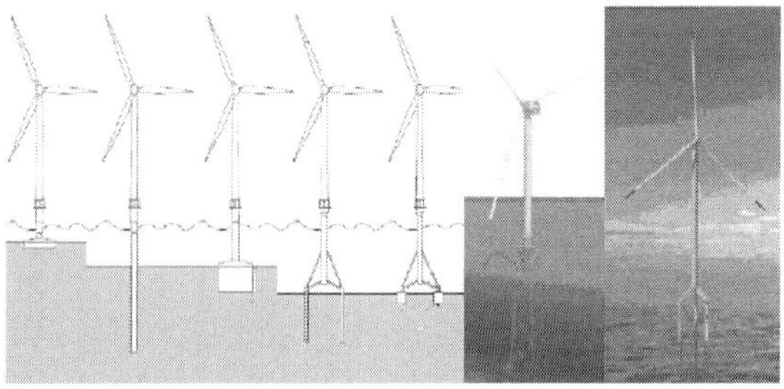

Figure 5.4: Offshore wind energy.

Source: Image by Shutterstock.

5.4. HISTORY OF WIND ENERGY APPLICATIONS

The use of wind energy can be traced from the last thousands of years to a number of ancient civilizations. It is revealed by the ancient human history that wind energy was discovered and used independently at various sites of the earth (Figure 5.5).

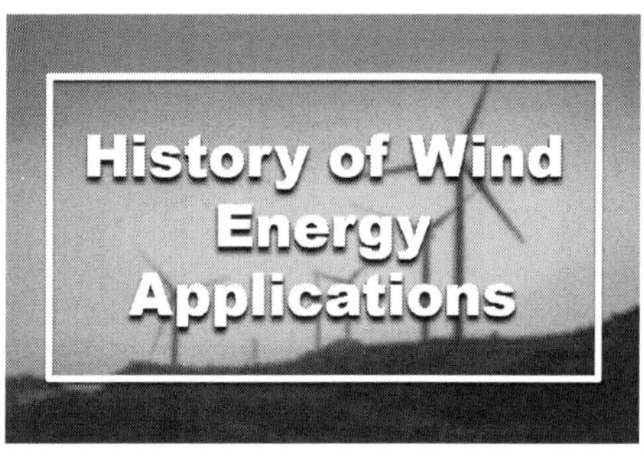

Figure 5.5: History of wind energy applications.

5.4.1. Sailing

The ancient Chinese were the first to attach sails to their primitive rafts, as early as about 4000 B.C. The ancient Chinese scripted on turtle shells from the oracle bone inscription in Shang Dynasty (1600 B.C.–1046 B.C.). Chinese junks were developed and used as ocean-going vessels in Han Dynasty (220 B.C.–200 A.D.).

According to the information recorded in a book written in the third century, there were multi-mast, multi-sail junks sailing in the South Sea, and they were capable of carrying 700 people with 260 tons of cargo.

The famous encyclopedic science and technology book Exploitation of the works of nature quoted that two ancient Chinese junks shown is a two-mast Chinese junk ship for shipping grain. In Song Dynasty (960–1279), it illustrates a wheel boat. This sort of wheel boat was used at the time of the war between the Jin Dynasty and Song (1115–1234).

Initially, the ancient Egyptians launched their first sailing vessels to sail on the Nile River and later along the coasts of the Mediterranean approximately at 3400 BC. Egyptians built sophisticated ships to sail on the Red Sea in 1250 BC. For a long time, wind-powered ships had dominated water transport until the invention of steam engines took place in the 19th century.

5.4.2. Wind in Metal Smelting Processes

About 300 BC, to provide furnaces with sufficient air for raising the temperatures inside furnaces in excess of 1100°C in iron smelting processes, ancient Sinhalese had taken advantage of the strong monsoon winds. This technique was capable enough for bringing high carbon steel.

In the fourth century BC, the double-acting piston bellows were invented in China and were widely used in metallurgy. To raise high enough temperatures for smelting iron, it was the capacity of this type of bellows to deliver continuous blasts of air into furnaces. Ancient Chinese could once cast many tons of iron in such a way.

5.4.3. Windmills

China is using windmills for a long time. At Sandaohao, Liaoyang City, the unearthed mural paintings from the tombs of the late Eastern Han Dynasty (25–220 AD) have shown the exquisite images of windmills that provide evidence of the use of windmills in China for at least approximately 1800 years.

In Sistan (eastern Persia), as recorded by a Persian geographer in the ninth century, the practical vertical axis windmills were built for grain grinding as well as water pumping. In the 1180s, the horizontal axis windmills were invented in north Western Europe.

Earlier, the windmills exhibited four blades and were firmly fixed on central posts known as Post mill. Later, based on the improvements on Post mill, a number of types of windmills had been developed in the Netherlands and Denmark, for example, Dutch mill, Smock mill, and Fan mill. In Europe and North America, the horizontal axis windmills have become dominant for several centuries because of their higher operation efficiency as well as technical advantages over vertical axis windmills.

5.4.4. Wind Turbines

Wind turbines are used to convert wind energy to electricity, unlike windmills that are used directly to do work like grain grinding or water pumping. In the year 1888, Charles Brush built and designed the first automatically operated wind turbine in the world (Figure 5.6).

Figure 5.6: Wind turbines.

Source: Image by nature.

During that time, the wind turbine was equipped with 144 cedar blades having a rotating diameter of 17 m. It generated a peak power of 12 kW to charge batteries that supply DC to electric motors as well as lamps. In the mid-1950s, the Gedser wind turbine was built in Denmark as a pioneering design for modern wind turbines. In the present world, modern wind turbines in wind farms consist of three blades that operate at relatively high wind speeds for the power output up to several megawatts.

5.4.5. Kites

Kites were invented as early as the fourth or 5th centuries BC in China. Han Fei-Zi (280–232 BC), a famous Chinese ancient legalist, mentioned in his book that an ancient philosopher Mo Ze (479–381 BC) spent 3 years to make a kite by using wood; however, he was not successful after the one-day flight.

5.5. ENVIRONMENTAL IMPACT OF WIND ENERGY

Reduction in the usage of fossil fuels, leading to a reduction in emissions of pollutants created by combustion leads to the environmental benefits of wind power. Perhaps, the absence of gaseous emissions, such as COx, SOx and NOx is the most important of these. Wind plant does not contribute to the greenhouse gas effect because it does not emit CO_2.

Many conventional fossil fuels generating stations emit NOx together with SOx. These stations have led to the acid rain effect and as a result, have caused significant environmental damage. The data involves direct as well as indirect emissions. The precise investment in gaseous emissions relies upon the mix of generating plants in the power system with which the wind farm is related. During the manufacture of a wind turbine plant, any emissions caused are offset after a few months of emission-free operation.

In the same way, after about a year's operation, the energy expended in manufacturing a wind turbine is paid back (WEC, 1994). The environmental damage of wind turbines involves land use, impact on local ecology, visual intrusion, noise as well as effects on radio communications together with television reception.

Despite that, in many cases, they can be alleviated by a policy designed to secure the maximum public acceptance of wind energy and by careful siting and design of wind plants. Usually, the most frequent public objection to wind turbines is a visual intrusion on the landscape.

The windiest sites have the least sign of human interference and are often the most beautiful. Generally, the response of the public towards visual impact is conditioned by attitudes to various types of energy production and the merits or demerits people see for themselves as well as their society. Demonstrate properly sited, well-maintained wind turbines that operate efficiently are the best ways to present the benefits of wind power.

According to the views of Walker and Jenkins (1997), the height, size, material, color, wind farm configuration, landscape background, and sitting to reduce lines of sight are all design factors that should be taken into consideration to decrease visual impact.

Dual land use is one of the merits of wind farms. Energy of wind is diffuse and capturing energy at a large scale from the wind requires use of turbines over a wide area.

In the year 1993, Grubb, and Meyer observed that modern wind farm uses only 1% of the total land area it occupies and the wind turbines together with their bases occupy only 0.2% of the total land area. Provided they are properly planned, the left-over land can be used for agriculture purposes and often wind farm access roads are seen by farmers to be an asset instead of a disadvantage. The level of possible disturbance to the local ecology is the result of both construction as well as operation of projects related to wind energy is a concern of a number of people. As per the conventional power station standards, the period of construction for a wind farm is very short by and is typically less than six months.

The process throws a small proportion of the land area into disorder designated for the plant. However, later on, the site can be returned to its former condition except for the small area required for the turbines and access roads.

In the year 1993, Benner surveyed across several countries that the impact of wind turbines on birdlife is an important matter. Due to wind turbines, birds can be affected in two ways: by death or by injury following a collision with the tower or blades, or by being disturbed in their nesting, breeding, or habits of feeding. In extreme situations, victims of birds can number from 0 to several 100 per year.

The numbers of victims of birds are often less as compared to those affected by equivalent lengths of high voltage lines. Clausager conducted a study in the year 1996 and concluded that for the 3500 wind turbines in Denmark, some 20,000–25,000 birds die yearly due to collision with wind turbines, on the other hand, over a million birds lose their life by traffic every year. The disturbance to breeding birds is more serious to resting and migratory birds but appears negligible. There is a need to avoid sitting of wind turbines on migration routes and where there are high densities of nesting or foraging birds. Careful and detailed design of the towers as well as turbines can help in reducing the nesting and perching opportunities for birds.

5.6. WIND GENERATION

The movement of air because of atmospheric pressure gradients results in wind. Wind flows from areas of high pressure towards t areas of low pressure. The higher the wind speed, the larger the atmospheric pressure gradient and thus, the greater the wind power that can be captured from the wind by means of wind energy-converting machinery.

Due to a number of factors, the generation and movement of wind are complicated. The most important factors among them are uneven solar heating, the Coriolis effect due to self-rotation of the earth as well as local geographical conditions.

5.6.1. Uneven Solar Heating

The uneven solar radiation on the surface of the earth is the most important and critical factor among all factors affecting wind generation. There are four reasons that can be attributed to unevenness of the solar radiation. The first reason is that the earth is a sphere that revolves around the sun in the same plane as its equator.

The equator receives the greatest amount of energy per unit area because the surface of the earth is perpendicular to the path of the sunrays at the equator but parallel to the sunrays at the poles, with energy dropping off toward the poles.

It forms a temperature gradient from the equator to the poles and a pressure gradient from the poles to the equator due to the spatial uneven heating on the earth. Thus, cold air with higher density flows from the poles towards the equator along the surface of the earth and hot air with lower air density at the equator rises to the high atmosphere and moves towards the poles.

The air circulation at each hemisphere forms a single cell without considering the earth's self-rotation and the rotation-induced Coriolis force, defined as the meridional circulation. Second, the self-rotating axis of the earth has a tilt of about 23.5° in the context of its ecliptic plane. During the revolution around the sun that results in cyclic uneven heating, it is the tilt axis of the earth causing the yearly cycle of seasonal weather changes.

Third, the surface of the earth is covered with several different types of materials, for example, sand, vegetation, water, rock, ice, or snow, etc. Each of these materials has different absorbing as well as reflecting rates to solar radiation even at the same latitudes that lead to high temperature in some areas (such as deserts) and low temperature in others (such as iced lakes).

The fourth and last reason behind uneven heating of solar radiation are the topographic surface of the earth. There are a large number of valleys, hills, mountains, etc., on the earth that give rise to different solar radiation on the sunny as well as shady sides.

5.6.2. Coriolis Force

Another important factor to affect wind direction and speed is the self-rotation of the earth. The force generated from the self-rotation of the earth is Coriolis force. This force deflects the direction of atmospheric movements (Figure 5.7).

Figure 5.7: Coriolis energy.

Source: Image by renewables now.

In the south atmosphere, the wind is deflected to the left, wherein the north atmosphere wind is deflected to the right. Latitude of the earth provides the basis for the Coriolis force. It is zero at the equator and ends up at maximum figures at the poles. Over and above, wind speed also provides the basis for deflection on wind; stronger blowing wind is deflected more amounts, while slowly blowing wind deflected only a small amount.

In large-scale atmospheric movements, the Coriolis force due to the earth's self-rotation and the combination of the pressure gradient due to the uneven solar radiation causes the single meridional cell to break up into three convectional cells in each hemisphere: the Ferrell cell, the Hadley cell, and the Polar cell. Each and every cell has its characteristic pattern of circulation.

The Hadley cell circulation lies between the equator and north latitude 30°, dominating tropical along with sub-tropical climates in the Northern Hemisphere. In the upper atmosphere, the hot air rises at the equator and flows toward the North Pole. Coriolis force deflects this moving air to create the northeast trade winds. Coriolis force becomes so strong to balance the pressure gradient force at approximately north latitude 30°. On the grounds of this, the winds defect to the west direction.

The air collected at the upper atmosphere gives rise to the subtropical high-pressure belt. Therefore, sinks back to the surface of the earth, splitting into two components: one move along the surface of the earth toward North Pole to form the Ferrell Cell circulation that lies between north latitude 30° and 60° and another return to the equator to close the loop of the Hadley cell.

Until it collides with the cold air flowing from the North Pole at approximately north latitude 60°, the air circulates toward the North Pole along the surface of the earth. In this zone, the moving air is deflected to produce westerlies under the influence of Coriolis force.

The Polar cell circulation has its existence in the middle of the North Pole as well as north latitude 60°. At the North Pole, the cold air sinks down and flows along the surface of the earth toward the equator. The Coriolis Effect plays an important role in forcing the airflow to southwest near north latitude 60°.

5.6.3. Local Geography

A result of both natural geography and manmade structures is the roughness on the surface of the earth. Generally, frictional drag as well as obstructions near-surface of the earth retard with wind speed and, as a result, induce a phenomenon known as wind shear.

With the greatest rates of increases observed over the roughest terrain, the rate at which wind speed increases with height depends on local conditions of the terrain, topography, and climate. A reliable approximation is that speed of wind shoots up approximately 10% with each doubling of height. Over and above, a few special geographic structures can enhance the wind intensity in all respects. For example, the wind that blows through mountain passes with high speeds can form mountain jets.

5.7. INTEGRATION OF WIND AND OTHER ENERGY SOURCES

One of the notable characteristics in wind power generation, especially for off-grid wind power systems is its uncertainty because of the sudden change in both direction and wind speed. Hence, the power output from wind turbines keeps fluctuating from time to time (Figure 5.8).

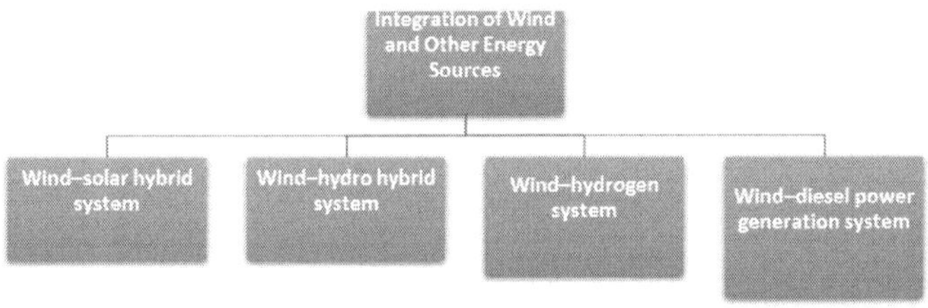

Figure 5.8: Integration of wind and other energy sources.

At the time of connecting wind turbines to an isolated or small grid, the power output from other generators must be varied to keep system frequency as well as voltage within predefined limits in response to these variations together with fluctuation.

It is beneficial to integrate wind and all other complementary energy sources to form hybrid power systems for this purpose to assure the reliability and stability of the power supply and reducing the need for the storage of wind energy.

5.7.1. Wind-Solar Hybrid System

Solar as well as wind energy are indeed high source of intermittent electricity generation. Time intervals within which fluctuations take place span a number of temporal scales from the period of seconds to years. These fluctuations can be further subdivided into periodic fluctuations (annual or diurnal fluctuations) as well as non-periodic fluctuations as per the change in weather.

Solar and wind energy are interrelated in time sequence as well as regions. In summer, sunlight is intensive and the sunshine duration is comparatively long however, the wind is less. In winter, wind becomes strong when less sunlight is available. The sunshine is strong on the other hand; the wind is weak during the day.

The wind is strengthened because of large temperature changes near-surface of the earth after the time of sunset. In Serbia, a study revealed that the effects of complementarity are more dramatic in certain locations and periods.

The optimum combination of the wind-solar hybrid system ranges in the middle of 0.70 and 0.75 of solar energy to load ratio with the minimized life cycle cost as per the analyzes and test data of wind-solar hybrid power systems.

The leveled energy cost for the wind-solar hybrid system is always low as compared to that of standalone wind or solar system for all load demands. Wind-solar hybrid power systems can ensure the reliability of electricity supply as the major operating time for wind and solar systems occur at different times.

The applications related to wind-solar hybrid systems range from residential houses to municipal as well as industrial facilities, standalone or grid-connected. For example, wind-solar hybrid systems as an independent power supply source have been widely used for street lighting purposes in China.

The largest wind-solar power test base of the world that integrates wind power, PV power as well as energy storage is located at Zhangbei, China. The project will have an installed capacity to generate 75 MW of chemical energy storage, 100 MW of solar power and 300 MW of wind power.

5.7.2. Wind-Hydro Hybrid System

Hydropower generation involves the conversion of potential energy in water into electrical energy with the usage of hydropower generators. Hydropower is renewable in nature. As it is also a clean energy source, accounts for the dominant portion of electricity generated from all renewable sources. In various locations of the world, hydropower is complementary to wind power (Figure 5.9).

Figure 5.9: Wind-hydro hybrid system.

It is lower in the dry seasons (winter and spring) but higher in the wet seasons (summer and fall), whereas the seasonal wind power distribution is higher in winter and spring but lower in summer and fall.

Hence, the integration of wind as well as hydropower systems is capable of providing significant economic, technical, and systematic advantages for both systems. "Green" electricity can be produced with wind-hydro hybrid systems taking a reservoir as a means of energy regulation.

5.7.3. Wind-Hydrogen System

Hydrogen is an energy carrier. It requires a number of resources like biomass, fossil fuels, and water for its production. As it has been known that hydrogen is a fuel with high energy density, therefore it can be transported, stored, and then converted into electricity through the medium of fuel cells at end users.

Generally, it is accepted that solar power, wind power as well as various other renewable energy power generation systems can be coordinated well with the electrolysis hydrogen production system to produce hydrogen fuel.

In the UK, the largest wind hydrogen power system has been applied to a building that is fueled solely by wind and "green" hydrogen power with the developed technology of hydrogen mini-grid system. As per this system, electricity generated through wind turbine is firstly provided to the building and later on excess electricity is used to produce hydrogen with the help of a state-of-the-art high-pressure alkaline electrolyze.

5.7.4. Wind-Diesel Power Generation System

To provide a stable supply of electricity, wind power can be combined with power produced by diesel engine-generator systems. Diesel generator sets may operate intermittently to reduce the consumption of the fuel in response to the variations in wind power generation as well as electricity consumption.

A study revealed that in comparison with power supply from diesel generation alone, the wind-diesel standalone systems can operate with an estimated 50–80% fuel saving. Since 1995, wind-diesel hybrid power systems have been studied in the US.

A large number of wind-diesel power generation systems have been installed all over the world, and a number of new techniques have been developed. Three different types of wind-diesel systems can be distinguished as per the proportion of wind use in the system: low, medium, as well as high penetration wind-diesel systems.

Presently, high penetration wind-diesel systems are at the demonstration level, whereas low penetration systems are used at the commercial level. The trends of technology involve the advancement of robust together with proven control strategies.

5.8. GRIDS, MARKETS, AND SYSTEM INTEGRATION

The introduction of significant amounts of wind energy into the power system involves a series of economic impacts that are both positive as well as negative in nature. Two main aspects determine wind energy integration costs at the power system level. These two aspects are balancing needs and grid infrastructure.

The inherently variable nature of wind power gives rise to the additional balancing cost in a power system. It requires changes in the configuration, scheduling together with operation of other generators to deal with unpredicted deviations between demand as well as supply (Figure 5.10).

Figure 5.10: Grids, markets, and system integration.

Here, it has been demonstrated that sufficient evidence is available from national studies to make a good estimate of such costs. Moreover, it also shows that they are low as compared to the overall balancing costs of the power system and with the generation costs of wind energy.

It is necessary to upgrade the network for a number of reasons. There is a need to provide additional transmission lines and capacity to reach and connect present as well as future sites of wind farms and to transport power flows in the distribution together with transmission networks.

The outcome of these flows emerges both from increasing demand as well as the trade of electricity and from the rise of wind power. Depending on the technical characteristics of the wind projects and trade flows, the networks must be adapted to upgrade voltage management at significant levels of wind energy penetration.

Moreover, the restricted interconnection capacity typically refers to the advantages arising from the omnipresent, widespread nature of wind; many other renewable energy sources and electricity trade in general are lost. In this regard, any improvement in infrastructure results in multiple benefits to the whole system.

Due to this reason, its cost should not be allocated to wind power generation only. As the proportion of wind power rises, the cost of modifying the power system increases in a more or less linear way, and it is not easy to identify its 'economic optimum' as costs are accompanied by advantages.

As per the studies done so far, and by implementing their results to higher wind energy penetration levels, one can easily conclude that it is economically (as well as environmentally) beneficial to integrate over 20% wind power into the EU power system.

Furthermore, a 20% share of wind power of EU electricity demand is not an upper limit. However, a number of benefits of wind energy must be considered, which include the contribution that it makes to the environment, security of supply and the other benefits.

5.8.1. Grid Losses, Grid Reinforcement, and Grid Management

Generally, wind power is induced in remote locations of the electricity grid. In simple terms, wind power may contribute to eliminating grid losses. On the contrary, wind farms may also be settled in areas that are remote and have low population density as well as a weak electrical grid consequently (Figure 5.11).

Figure 5.11: Grid losses, grid reinforcement and grid management.

This may result in extra costs for reinforcement of the regional transmission grid (generally fewer than 400 kV) and possibly the bulk transmission grid (generally above 400 kV). Additionally, in regard to stabilizing the grid, serial electrical compensation equipment may be required (depending on characteristics of the grid together with the electrical properties of the wind turbines).

5.8.2. Intelligent Grid Management

The development of wind energy is facing an important limitation and that is obstacles in the electrical transmission grid internationally. One of the reasons is that superfine wind resources (for example, gas, oil, and coal) are often located in remote areas that are less populated and are under developed. They have (thermally) limited transmission capacity to other parts of the electrical grid where electricity can be utilized.

Minimization of cost per kWh usually implies a capacity factor of around 30% for a given wind climate. Since wind speeds statistically follow a skewed distribution, low wind speeds are very frequent, whereas high wind speeds occur only very rarely.

In simple words, in case electrical interconnections are dimensioned to meet the maximum power output of wind farms, they will be start getting used inefficiently. Furthermore, the peak production of wind farms will rarely coincide when several are sufficiently geographically dispersed within a transmission-congested area.

At last, wind power generation often fits well with local power demand in temperate climates. This will certainly diminish the amount of transmission capacity that is needed to some extent.

5.8.3. Cost of Ancillary Services Other Than Balancing Power

A term, ancillary services that are often used for a number of safety mechanisms that have been built into the generating units in the electrical grid. These services provide stabilizing mechanisms or ensure an efficient transfer of energy by using the grid (in the case of reactive power control) which serve to avoid catastrophic grid collapse (blackout) so that errors in a single grid component or influence from lightning strikes do not cascade by the electrical grid (Figure 5.12).

Figure 5.12: Cost of ancillary services other than balancing power.

In the past, wind turbines were designed as passive components when they were only intended to provide a small part of total generation, in other words, the turbine would cut itself off from the grid and stop turning in case a wind turbine control system detected that grid frequency or grid voltage was outside a permitted range.

With large amounts of wind power on the grid, this is not an appropriate reaction in the case of a collapse in voltage since it may exacerbate the initial grid instability problem. Therefore, modern wind turbines are designed as active grid components. It is able to contribute to stabilizing the grid in case of electrical grid errors.

This is the case for voltage and frequency control, reactive power control, as well as 'fault ride though' capabilities of wind turbines. The costs of these features are usually included in the turbine price that meets the local grid connection requirements.

5.8.4. Providing Balancing Power to Cope with Wind Variability

Since the other turbines in the grid will largely cancel out the variations, minute to minute or second to second variations in wind energy production is rarely a problem for installing wind power in the grid. However, the production of wind turbine energy may vary from hour to hour, just as electricity demand from electricity customers will vary from hour to hour.

This means that other generators on the grid have to provide power at short notice to balance demand as well as supply on the grid in both cases. The cost of providing this balancing service depends both on the predictability of the variation in net electricity demand that is demand variations minus wind power generation and on the type of other generating equipment available on the grid.

The more predictable the net balancing needs, the easier it will be to use the least expensive units to provide the balancing service and the easier it will be to schedule the use of balancing power plants (in simple words, to regulate generation up and down at short notice period).

As mentioned previously, wind generation will generally reduce the hour-to-hour variability of net electricity demand compared to a situation with no wind power on the grid because wind generation in temperate climates usually fits well with electricity demand.

As it is reliable to forecast wind generation a few hours ahead, thus, the scheduling process can be eased, and balancing costs can be lowered. A great number of commercial wind forecasting products are available on the markets that are often combined with upgraded meteorological analysis tools.

5.9. CONCLUSION

In the end, it is concluded that wind energy has been in existence for several decades. Its acceptance all over the world has been increasing rapidly. As it is low cost and its capacity of producing energy has been increased with advancements in technology, more people are installing it for a consistent supply of energy.

It is best applicable in areas where there is a strong wind, but usually has applicability in rural areas because of large areas and restricted power supply.

Nowadays, there are several types of wind energy systems available in the market, before choosing a particular one, it is essential to first look for several factors such as cost factors, land factors, and other requirements for long term success.

REFERENCES

1. Paul Gardner, Andrew Garrad, Lars Falbe Hansen, Peter Jamieson, Colin Morgan, Fatma Murray, (2021).*Wind Energy-the Facts*. [eBook] European Wind Energy Association. Available at: https://www.wind-energy-the-facts.org/images/chapter1.pdf (accessed on 5 May 2021).
2. Aliprantis, D., (2021). *Advantages and Challenges of Wind Energy*. [Online] Energy.gov. Available at: https://www.energy.gov/eere/wind/advantages-and-challenges-wind-energy (accessed on 5 May 2021).
3. Cox, D. L., (2003). *Wind Energy*. [eBook] Available at: https://www.aps.org/policy/reports/occasional/upload/wind_energy.pdf (accessed on 5 May 2021).
4. Encyclopedia Britannica. (2021). *Wind Power | Capacity and Facts*. [Online] Available at: https://www.britannica.com/science/wind-power (accessed on 5 May 2021).
5. Energy Matters, (2021). *How a Wind Turbine Works*? [Online] Available at: https://www.energymatters.com.au/components/wind-energy/ (accessed on 5 May 2021).
6. Fares, R., (2021). *Wind Energy is one of the Cheapest Sources of Electricity, and it is Getting Cheaper*. [Online] Scientific American Blog Network. Available at: https://blogs.scientificamerican.com/plugged-in/wind-energy-is-one-of-the-cheapest-sources-of-electricity-and-its-getting-cheaper/ (accessed on 5 May 2021).
7. Ge.com. (2021). *What is Wind Energy | GE Renewable Energy*. [Online] Available at: https://www.ge.com/renewableenergy/wind-energy/what-is-wind-energy (accessed on 5 May 2021).
8. Iberdrola, (2021). *What is Wind Energy? How is it Converted into Electricity and What are its Advantages?* [Online] Available at: https://www.iberdrola.com/environment/renewables-energy-wind-power (accessed on 5 May 2021).
9. Irena.org. (n.d.).*Wind*. [Online] Available at: https://www.irena.org/wind (accessed on 5 May 2021).
10. Kaygusuz, K., (2004). Wind energy: Progress and potential.*Energy Sources,26*(2), 95–105.
11. Krohn, S., Morthorst, P., & Awerbuch, S., (2021). *The Economics of Wind Energy*. [eBook] European Wind Energy Association. Available at: https://www.ewea.org/fileadmin/files/library/publications/reports/Economics_of_Wind_Energy.pdf (accessed on 5 May 2021).

12. National Geographic Society, (2012). *Wind Energy*. [Online] Available at: https://www.nationalgeographic.org/encyclopedia/wind-energy/ (accessed on 5 May 2021).
13. Nrel.gov. (n.d.).*Wind Energy Basics*. [Online] Available at: https://www.nrel.gov/research/re-wind.html (accessed on 5 May 2021).
14. Renewables First-The Hydro and Wind Company, (2021). *Characteristics of a Good wind Power Site-Renewables First*. [Online] Available at: https://www.renewablesfirst.co.uk/windpower/windpower-learning-centre/what-makes-a-good-wind-power-site/ (accessed on 5 May 2021).
15. Sorensen, B., (1995). History of, and recent progress in, wind-energy Utilization.*Annual Review of Energy and the Environment,20*(1), 387–424.
16. Tong, W., (2010). Fundamentals of wind energy.*Wind Power Generation and Wind Turbine Design,* 3–48.
17. Toshiba-energy.com. (2021). *Wind Power Generation Using Wind Energy : Systems and SolutionsRenewable Energy | Toshiba Energy Systems and Solutions Corporation*. [Online] Available at: https://www.toshiba-energy.com/en/renewable-energy/product/wind-power.htm (accessed on 5 May 2021).
18. Wagner, H., (2013). Introduction to wind energy systems.*EPJ Web of Conferences,54*, 01011.
19. Wagner, H., (2017). Introduction to wind energy systems.*EPJ Web of Conferences, 148*, 00011.

Chapter 6

Applications of Wind Energy in Agriculture

CONTENTS

Wind energy is being harnessed by mankind since the earliest civilization. Wind power is being used since centuries on farms in various activities, from grinding grains to flour and pumping water from deep wells and power sailboats. However, in recent years support for renewable energy, particularly, wind power has increased. Wind energy has time and again proved to be an excellent alternative form of energy, and its continual and successful use on farms for decades is proof of it.

The current chapter proposes to explain how its application will help optimize agriculture by bringing down costs and promoting clean and renewable energy. It will also discuss the history of the usage of wind power in agricultural practices, its environmental impact, its economic feasibility, and its shortcomings, amongst other topics.

6.1. INTRODUCTION

Wind energy is the use of wind to provide automatic energy through wind turbines to turn electric generators for electrical power. Wind power is a highly sought-after sustainable and renewable source of energy that lessens the implications on the environment when created and harnessed in comparison to the burning of fossil fuels.

Agriculture is the only source of human food. Of late, the surging demand for food and the unstoppable price hike of fossil fuels and the sheer damaging environmental hazards caused by conventional agricultural techniques, amongst many other factors are leading to a rise in the adoption of renewable sources of energy to mitigate any further damage.

Energy contributes as one of the largest overhead costs in agricultural production and is one of the major parameters for establishing growth in it. In the current scenario, the challenges agriculture faces are modernization and efficiency as also reduction of the rising energy costs to drive it. Machines used on farms are primarily functioning on fossil fuels, which in turn result in the emission of greenhouse gases and irreversible climate change (Figure 6.1).

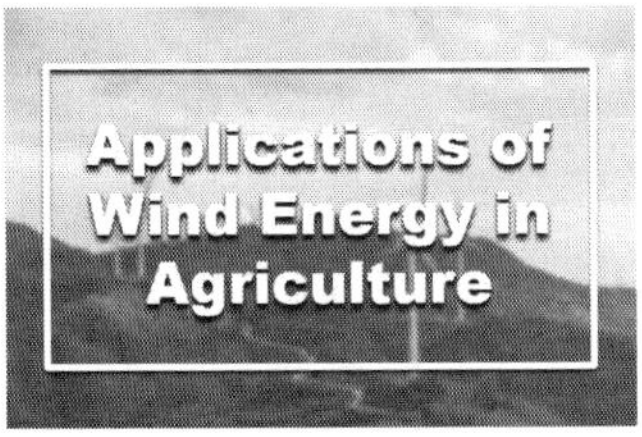

Figure 6.1: Application of wind energy in agriculture.

There are a variety of agricultural activities that require huge amounts of energy such as irrigation, land preparation and fertilization, livestock rearing operations, etc., these aforementioned activities coupled with many others can explain the dire need for the promotion and adoption of renewable energy resources in agriculture.

In agricultural systems, energy sources that release available energy directly to the system are classified as direct energy sources. Wind, solar, and biomass energy can be harvested forever and thus qualify as direct renewable energy sources. This chapter is an overview of the application and uses of wind energy for agriculture.

6.2. DEFINITION AND HISTORICAL BACKGROUND

Wind energy is a form of solar energy. The sun's energy creates temperature differences that drive air circulation. Hot air rises, reducing the local atmospheric pressure; nearby cooler air flows into this region of lower pressure; this airflow is wind (Figure 6.2).

Figure 6.2: Wind turbines in action on a farm.

Source: Image by windsystemsmag.com.

Wind energy is the kinetic energy of air in motion, also called wind. Wind turbines are used to convert the kinetic energy present in wind into mechanical power. A generator can further convert this power into electricity. Uneven heating of the atmosphere by the sun, variations in the earth's surface, and rotation of the earth are the causative factors of wind.

Wind flow patterns are influenced by both local and global phenomena. The topography of the region, such as mountains, water bodies, and flora or vegetation determines different wind directions. The energy in wind is transformed into electricity by the rotating propeller-like blades around a rotor.

The rotor turns the driveshaft, which turns an electric generator. The speed of the wind, air density, and swept area are the key factors that impact the amount of energy a turbine can harness from the wind.

The equation for wind power:

$$P = \frac{1}{2}\rho A V^3$$

6.2.1. Wind Speed

The amount of energy in the wind varies with the cube of the wind speed, in other words, if the wind speed doubles, there is 8 times more energy in the wind ($2^3 = 2 \times 2 \times 2 = 8$). Small changes in wind speed have a large impact on the amount of power available in the wind.

6.2.2. Density of the Air

The denser the air, the more energy received by the turbine. Air density varies with elevation and temperature. Air is less dense at higher elevations than at sea level, and warm air is less dense than cold air. All else being equal, turbines will produce more power at lower elevations and in locations with cooler average temperatures.

6.2.3. Swept Area of the Turbine

The larger the swept area (the size of the area through which the rotor spins), the more power the turbine can capture from the wind. Since the swept area is $A = pir^2$, where r = radius of the rotor, a small increase in blade length results in a larger increase in the power available to the turbine.

The use of wind power can be dated back to the prehistoric times when humans have put sails into the wind and travelled the world. For more than two millennia now, wind-powered machines have ground grain and pumped water. Wind power was widely available and not confined to the banks of fast-flowing streams, or later, requiring sources of fuel.

The windmill and wind pump were first developed in present-day Iran, Afghanistan, and Pakistan around the 9th century. Slowly it became widespread practice across the Middle East and Central Asia and later spread to China and India. Later, vertical windmills became more common across North-western Europe.

The late middle ages saw the extensive spread of windmills across Europe, and by the late 18th century windmills entered America during the American Revolution. The 19th century saw a revolutionary landmark with the first wind turbine being used for the production of electricity in Scotland in July 1887 by Prof James Blyth of Glasgow.

By 1900, Denmark had around 2500 windmills used for mechanical loads such as pumps and mills. From 1850 to 1900, the American Midwest witnessed about 6 million small windmills, being installed on farms to run irrigation pumps.

The 20thcentury saw the emergence of wind energy as a competitor against traditional fossil fuels due to the surging oil prices and in the 21'st century, fossil fuels being comparatively cheaper yet the burning issues of energy security, global warming, and the ultimate fossil fuel exhaustion led to a steady increase in preference of renewable forms of energy.

6.3. APPLICATION OF WIND ENERGY IN AGRICULTURE

Wind technology provides both mechanical and electrical energy for the process of agriculture. Wind power has been advancing at a rapid rate and has overtaken biopower to stand out as a preferred form of energy. The minimization of greenhouse gas emissions can be brought about by the production of eco-friendly power generation technologies such as wind energy.

This requires undertaking major research in environmentally friendly technologies. Shifting to clean energy sources will help reduce polluting energy production. Fossil fuels are being predicted to be on the verge of running out completely by the end of this century. Additionally, carbon emissions continue to rise, causing major environmental problems. Thus, wind energy is the best alternative energy source as it is the need of the hour.

Amongst all renewable energy resources, wind energy is clean and simple to be harvested. Its application in agriculture dates back centuries ago and its newfound industrial and household usage has made it an even more popular choice today.

A report in 2011 from the Global Wind Energy Council states that the electricity generated by wind energy in 2010 had increased 35,802 MW which was 22.5% of the total 194,390 Mega-Watt electricity made by wind energy in the world.

Half of these were produced in China, where the economy was booming as well as the need of energy resources. In countries such as India, Japan, South Korea, and Taiwan, the use of wind power was also steadily increasing every year. Thus, Asia emerged as one of the growing regions in wind energy usage with the capacity of 19,022 Mega-Watt in 2010.

Wind energy is not only harvested to produce electricity but also for specific tasks related to agriculture such as water pumping, irrigation, irrigation, heating farm structures, product processing, cooling, and storage, and so on.

Very few generalizations can be derived from the careful study of the patterns of power and energy consumption and usage of the applications of wind energy in agriculture. In this chapter, each type of application shall be described individually to draw up the plan and framework using wind power which shall be both practical in its application and affordable at the same time.

This topic of the chapter further presents a list of objectives that must be met before the actual practical application of wind energy takes place.

6.3.1. General-Purpose Farmstead Electric Power

Interfacing a wind turbine with the electric power system on a farm would lead to a partial supply of the electrical energy used on the farmstead. In this case, it would so happen that the energy generated by the wind would be consumed in case the immediate direct farmstead load is more than the instantaneous wind-generated power with the remainder of the power being made up by the utility.

In case the instantaneous wind-generated energy has more capacity and exceeds the instantaneous farmstead load the excess wind-generated would flow back into the utility system.

It shall be noted that wind-generated electric power on the farm will significantly vary with time. Mostly, it can be seen that although the load on the farmstead is assorted, it will not usually be sufficiently diverse to smoothen out the total farmstead load (Figure 6.3).

Figure 6.3: Wind energy plays an important role in agriculture.

Thus, in situations where the energy generated from wind is not efficiently managed, a considerable amount of wind-generated energy will not be used on the farmstead but will flow to the utility lines (at a credit less than the retail rate), and, contrarily most of the electric power actually being put to use on the farmstead will come from the utility supply.

Thus, the farm owner of such a wind-powered generator will receive considerably less than the full economic value produced by the wind electric generator, and its installation has had substantially no effect on the utility's distribution costs since the expected peak demand would remain unchanged.

A thorough assessment of farmstead electric loads suggests that a significant proportion of these loads are installed on a device or system which has an in-built capacity for storage of work done. This feature would enable switching on the load to be deferred for a time after automatic controls would normally switch on the load.

These loads, for activities such as water heating, building heat, feed grinding often constitute a considerable amount of the energy used on many farms.

This could be achieved by keeping them switched off during periods when wind-generated power is insufficient to supply power for them. Additionally, storage capacity could also be enhanced further for the system associated with the deferred loads, for example, larger or additional tanks for hot water and larger bins for ground feed.

6.3.2. Irrigation Pumping

Water-pumping windmills, aero generators (small wind electric generators), and wind-solar hybrid systems have been deemed extremely useful for meeting water-pumping and small-power requirements in a decentralized model in rural and remote windy areas of the country. It requires the largest on-farm energy use for stationary applications in agriculture today (Figure 6.4).

Figure 6.4: Irrigation pumping in agriculture.

As a result, the substitution of natural gas or electricity wind power might substantially reduce the total fossil fuel usage in agricultural activities and the cost of operation for the farmer. The escalating costs of energy (natural gas or electric energy) are critically impacting the production decisions of farmers. Many farmers have abandoned irrigation and have resorted to dry land farming. Thus, it can be established that farmers are looking for cheaper and sustainable sources of pumping energy.

6.3.2.1. Case Study on Drip Irrigation System Using Wind Energy

The following case study will highlight automatic drip irrigation systems using wind energy. With drip irrigation, the water is conveyed under pressure through a pipe system with a small diameter and then slowly emitted into the soil.

The emitters are placed nearby the plants so water can reach the crop roots directly, which lowers the evaporation. Because of this, a smaller amount of water is needed than when using surface or sprinkle irrigation. The water is emitted at a low rate, of only about 2–20 liters per hour and emitter, and the emitters can be placed on the surface or underground (Figure 6.5).

Figure 6.5: Drip irrigation using wind energy.

Source: Image by cimmyt.org.

The Wind Mill is the main part of this system which converts the wind energy into an electrical energy. The windmill converts the wind energy into mechanical energy and thereafter, mechanical energy is converted into electrical energy. After that this electricity is used in the farm to run water pump, controller, and other parts of this system.

The moisture levels in the soil play a crucial role here. There are various ways to measure the moisture of the soil. We are going to measure the moisture level by the conductivity property of the moist soil. We all know that moist soil conducts electricity better than dry one.

And the impedance level of the dry soil is higher than the moist one. Another one is water level sensor for checking whether the water is sufficient or not. This sensor automatically generates a signal and gives it to the microcontroller for proper water to the plants.

6.3.2.2. Objective

The objective of this activity is to obtain electricity using a windmill and utilization this energy for the irrigation purpose being run automatically with the help of moisture sensors and a microcontroller. Utilization of wind energy by setting up a windmill followed by the storage of the acquired power is our work part. The stored power is to be utilized for the driving control unit, thus running an automatic pump-controlled irrigation system.

The objectives of this project are:

- To design, implement and provide a simple and economic solution for water lifting and irrigation.

- The Electromechanical system should be developed for rural area as it suffers from power supply failure by using nonconventional energy sources.
- This system should operate on 12-volt DC, which is less hazardous than the present high voltage system.
- The system should monitor water content in soil and should regulate water release.
- The system should generate sufficient power for operation of additional components.
- The system should be easily configurable by minor changes in hardware and software.
- To make the system more user friendly and greater reliability.

6.3.2.3. Methodology

Wind energy has been selected prime source for the generation of electricity. Now to fetch this energy source a windmill cum vertical turbine has been set up which would rotate due to the kinetic energy of the wind, i.e., the intensity of wind blow.

This turbine will be rotating and shall be connected to the power generator via a gearbox. Such an arrangement would resemble to the functioning of a dynamo which will convert the mechanical energy of the turbine fan to electrical energy.

As the wind energy source is discontinuous so to make the power supply independent of the continuous wind, we have preferred using a battery to store the power being generated instantly. This stored power can be further used at any time, even if there could be the absence of wind. Since the project is based on DC power, we have used a Li-ion battery.

However, on the commercial scale, we would be having the generation of ac power so the battery used here would be replaced by an inverter. Now further controlling includes automatic switching system controlled by the microcontroller control unit.

The incoming power supply shall be made dc using a regulated power supply. Next, the input ports of the controller have been connected to the soil moisture sensors TNST-45 which would sense the moisture level of the field soil. Right after sensing it will send an electrical signal to the microcontroller, and the controller would start the pump connected at its output port accordingly with respect to the sensor.

Water flows out of the pump through the drip pipeline system. In this way, the required field can be irrigated and leaving the nonrequired area unwatered automatically without man supervision.

6.3.2.4. Result and Discussion

The idea behind this project is to develop an inexpensive and user-friendly system that will provide an additional sense to the Agriculture field so that the expenses toward the Electricity, Manpower, and water requirement can be minimized and more efficient water delivery can be made possible. The controlling of the automatic watering system is used in a farm:

- On the basis of the analysis carried out in the present work, the following conclusions are drawn.
- Wind energy is used for providing the proper voltage to the microcontroller and water-lifting for efficient working.
- Cost of this project is less as compare to other systems because sensors for water level and soil conductivity are used.
- In this work problem of over-irrigation which is because of poor distribution or mismanagement of wastes water, and may lead to water pollution is solved.
- Deep drainage (from over-irrigation) may result in rising water labels which in some instances will lead to problems of irrigation salinity and the said problem is addressed in this work.

6.3.3. Heating

Heating requires the second-largest on-farm energy use for stationary applications in agriculture and consumes approximately 10% of the energy used in agricultural production. Heating for the farmhouse constitutes the largest portion of this energy use, followed by the heating required for poultry and allied brooding and farming operations.

The heating of water for domestic use and dairy sanitation is an important energy use on farms and accounts for one-third of the rural use of electric energy.

The basic energy sources for heating in a rural setup are propane and fuel oil. The supply of both of these energy sources is depleting, leading to uncertainty and an increased price. As a result of the hike in the price of fossil fuels, a large number of rural consumers are installing electric heating systems and thereby aggravating the peak electric demand on the utility during the winter season.

Thus, using wind energy as an alternative energy source for heating will render two benefits, i.e., it can reduce our usage of non-renewable energy sources to a minimum and also empower rural electric power suppliers to lend additional electric heating loads by reducing demands on the rural power distribution systems.

There is a major revival in the interest expressed in the wind as a source of power in recent years, especially due to the rapid increase in energy costs and the future prospects of acute energy shortages. Approximately 2% of solar energy that reaches the earth is converted into wind energy.

The heating requirements of a farmhouse are pretty much similar to those of urban residences. Apart from this advantage, rural locations also enable the installations of alternative energy sources such as windmills more easily as they offer more readily available land in contrast to urban-to-urban locations. Also, rural locations have other advantages such as fewer occurrences of problems with neighbors in relation to television, interference, noise, or esthetics.

With the gradual increase in liquefied petroleum (LP) gas and oil prices and their restricted availability, the electric power will most likely assume an increasing hold on the rural heating load. In case this rise in electrical heating load is faced with the use of resistance heating, electric furnaces with high peak electric demands or heat pumps, the peak loads on rural electric distribution will increase drastically. Power suppliers in rural areas will be badly hit and will face major challenges in keeping up with the energy requirements.

Wind energy can be applied in several ways for heating farm structures:

- The wind system will help conserve and prolong the depletion of non-renewable energy resources like oil or LP gas in a hybrid model with existing traditional heat sources. The system would be sized in a manner to completely avoid exceeding the required heating load of the building and would function in a way that the heat generated from the wind energy would only constitute a fraction of the total heating load supplied by the existing system and thereby the provision of storage would not be used.

- The wind system can be sized in such a manner that it contributes to a major portion of the heating load. The excess wind energy produced can be stored, in the form of heat, in a suitable medium like water. This stored energy would come in handy during phases of low wind energy to support the heating load, just as the stored heat present in other solar systems is put to use when weather conditions are not conducive towards solar heat collection.
- The portion of the heating load that is to be provided by the conventional energy resources would be correlated to the incremental economic return of a larger wind system.
- The wind energy setup can be designed in a way to support the entire heating load, and the energy that is remaining unused after consumption can be used or stored or converted for other purposes.

To meet up with the peak-power requirements on the rural electric systems, the creation of an energy store in combination with a wind power system would work wonders. The requirements can be met by supplying energy from both the wind-power system and the wind storage system to meet the maximum electrical demand.

Keeping in mind, the current high costs of constructing the new facilities to enable electric distribution capacity of the rural power systems, the value of the modification of peak power demands, and of load-leveling through the usage of a wind energy system with provision for storage of excess or unused energy may be equivalent or greater than the energy produced.

There is some link between heating requirements are wind speed, particularly in the types of construction used on farms. Therefore, a wind-powered heating system will provide more heat during periods when heat is needed the most, although it is still not known the extent to which this enhances the value of wind-powered heating.

The objectives of the application on wind-powered heating are:

- To evaluate the practicality of modifying heating systems for operation by wind power;
- To measure the performance of wind-powered heating systems;

- To determine the adequacy of procedures used for designing wind-powered heating systems;
- To determine any operating difficulties.

6.3.4. Water Pumping Windmill

The very foundation of wind power technologies is the transformation of wind kinetic energy to mechanical driving energy by some kind of rotor or windmill. There are many types of rotor construction for pumping and electricity generation. Local wind conditions, the water source and the aim of the water pumping are the determining factors for the given application.

Water-pumping windmills, aero generators (small wind electric generators), and wind-solar hybrid systems have been found to be functional for meeting water-pumping and small-power pre-requisites in a decentralized model in rural and remote windy areas of the country.

A water-pumping windmill pumps water from wells, ponds, and bore wells for drinking, minor irrigation, salt farming, fish farming, etc. Available windmills are of two types, namely direct drive, and gear type. The most commonly used windmill has a horizontal axis rotor of 3–5.5 m diameter, with 12–24 blades mounted on the top of a 10–20 m high mild steel tower. The rotor is coupled with a reciprocating pump of 50–150 mm diameter through a connecting rod.

Such windmills start lifting water when wind speed approaches 8–10 kilometers (km) per hour. Normally, a windmill is capable of pumping water in the range of 1000 to 8000 liters per hour, depending on the wind speed, the depth of water table, and the type of windmill. Windmills are capable of pumping water from depths of 60 m.

Since the fuel requirements are reduced to zero in their operation, water-pumping windmills have an extra advantage and preference as they can be easily installed in remote and windy areas which have no scope of conventional water pumping.

However, water-pumping windmills have limitations too. They can be operated satisfactorily only in medium wind regimes (12–18 km per hour). Further, special care is needed at the time of site selection as the sites should be free from obstacles such as buildings and trees in the surrounding areas. The costs involved in their installation and system are very high thus not economically a viable option for most individual users (Figure 6.6).

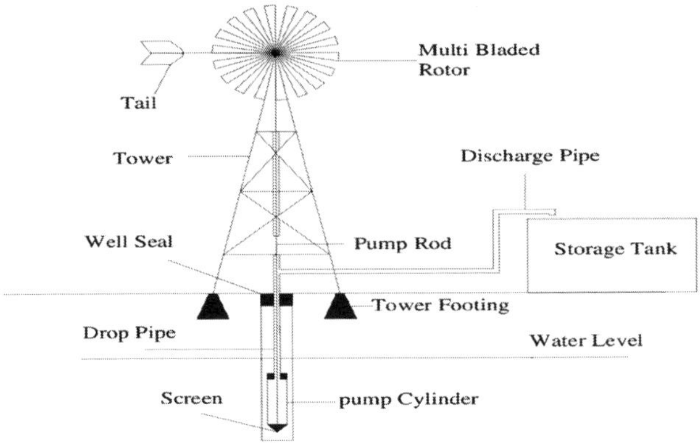

Figure 6.6: Water pumping windmill.

Source: Image by Agritech Tomorrow.

6.3.5. Information Dissemination

There has been comparatively less focus on the benefits and agricultural uses of wind power in any major federal publication in recent times. There have been privately published articles wrongly targeting wind energy and portraying it in a bad light as a clean source of energy.

The future will witness the agricultural sector adopting more clean renewable resources like wind energy, and much more research and the material will be introduced for better adoption and execution of the same. A variety of standard materials should be made available with great insight and detail into wind energy.

The objectives of the information dissemination program are as follows:

- To keep the agricultural sector and community well acquainted with the current Federal research program on the agricultural applications and usage of wind power.
- To keep both farmers and agricultural educators well-versed and well-informed about the criteria and benchmarks that should be kept in consideration while undertaking decisions on the application and use of wind power in agriculture.

For example, conduct workshops, hold information dissemination activities, etc., as these activities are solely aimed at early familiarization of the agricultural industry with the advantages and feasibility wind power holds and how it could be best used.

6.3.6. Wind Energy and Greenhouses

Applying wind energy to greenhouses is much more in spectrum than just a case of adapting to the power scale of a particular wind turbine. Turbine operation is possible only with the existence of a minimum number of wind hours; otherwise, it is highly unforeseeable with speeds above the threshold value for each device along with the possibility of spatial integration among them, especially in areas of thick greenhouse concentration.

The usable conversion power into mechanical energy, and subsequently into electrical energy, contained in the wind, can be measured by the following expression: Pw = 0.5*ρo*Aw*up(8.6) where Pw [W] is the power; ρ o [kg m^{-3}] is the air density; Aw [m^2] is the area swept by the wind turbine rotor; and up [m s^{-1}] is the wind velocity.

The non-linear nature of this expression shows the importance of adapting the conversion systems to a greater wind speed range. The energy conversion is, in any case, limited by the Betz limit, which establishes a 16/27 (~ 59%) ratio. Current wind turbine technology, whether on horizontal or vertical axes, allows for more intensive electricity production than other renewable technologies, such as solar power.

This is an interesting aspect in terms of its footprint, which is a critical aspect in greenhouse applications. To approximate, there is a 1:10 relationship in the power per unit of active area in each case (blades area/ module area). This determines that, under favorable operating conditions, the energy [kWh] produced by a wind turbine occupying the same space as a PV module will be substantially higher.

However, wind resource availability is strongly determined by factors such as climate and spatial factors (height, obstacles, and natural intensification pathways, etc.), which in turn influence its application and performance, future prospects and in this case also make it more restrictive.

6.3.7. Matching Crop Water Needs and Wind Energy

A wind turbine should be put to use throughout the year for economic purposes. Here, in the following case study, we assumed that the wind turbine was large; corresponding to farm-energy needs other than pumping, and was solely meant for irrigation pumping alone.

Thus, it was obligatory to study irrigation management practices and operational modes that would maximize the annual operating time and also replace a high percentage of fossil fuels currently used for pumping.

In the study carried out in Amarillo, Texas, Garden City, Kansas, and Columbus, Nebraska, we investigated the application of dedicated wind turbine systems without energy storage to irrigation pumping in the great plains and have come to the following conclusions:

- Using an auxiliary motor increases well yields in summer and therefore increases area to preplant irrigate per wind turbine.
- Fully irrigated summer crops with preplant irrigation will use 30 to 60% of the capturable wind energy, depending upon the crop and location.
- When the availability of water is restricted, and energy is costly (relative to returns), farmers often resort to limited irrigation of crops adapted to dryland farming. By doing so, 50 to 100% of the annual wind energy can be harnessed, depending upon the crops and management practices.

6.4. WIND ENERGY AND ITS IMPACT ON CROPS

New research in the field of wind energy indicates that it can do more than just improve the income from farms. The positioning and location of wind turbines in agricultural fields may help grower to grow crops better.

The research carefully studied America's famous Corn Belt which overlaps with its central "wind belt"—a wide belt of the midsection of the United States that is ideal for wind energy development.

The past decade has been witnessing incredible growth in wind energy with turbines often having been planted in or near cropland—leading both farmers and researchers to wonder and investigate what effect the rotating blades might have on corn, soy, and other crops (Figure 6.7).

Figure 6.7: Wind energy and its impact on the crops.

In conventional and old school farming styles and practices in many places, farmers grow trees along the edges of fields, a technique that slows the wind and stirs up the air, benefiting the crops in the field.

The possibility of turbines causing a similar effect on crops is being studied. Can turbines help crops to grow?

On-going research in the Midwest region of the U.S., which is touted as the heartland of the world's leading corn-producing nation, a region marked with stormy and windy fields providing an ideal location for wind energy farms. The findings here could apply to many places around the world, wherever turbines and farms are near each other, although the effects on vegetation may vary by region or by crop.

- **Mixing the Air:** One of the very apparent ways that wind turbines could help agriculture is by mixing up the air, getting more carbon dioxide (CO_2) to the crops, since "the job of corn is to take up as much CO_2 as it can," said Eugene Takle, an agricultural meteorologist at Iowa State University.

However, there are some more effects turbines could have on crops that are way more complicated. For example, "By causing the air to move, wind turbines could reduce the amount of dew on leaves at night. This would help reduce crop diseases, such as those caused by fungi. Takle said.

That would be a welcome impact in the top U.S. corn-producing state, Iowa, where climate change has made the air moister, making dew more common, according to a report this year commissioned by the Iowa state government. (Iowa was second to Texas in the United States in wind energy installed in 2011).

Another potential beneficial impact could be that since turbines mix up the air and slow wind speeds, they could also affect the temperature around them, making nights warmer and days cooler. Both these effects could help crops, making frosty nights less common, and reducing the number of hot days that bring stress to plants.

With this variety of effects, "we are finding it's more complicated than we thought," Takle said. But based on his work on how trees affect crops, Takle expects that "on balance, the effect would be positive."

- **Revealing Turbines'"Wakes":** To research further the impact turbines have on crops, Takle is collaborating with a team lead by renowned atmospheric scientist Julie Lundquist at the National Renewable Energy Laboratory (NREL) in Golden, Colorado.

Lundquist has begun using "lidar"—like radar, but using light—to reveal the swirls of air in the long "wake" behind a wind turbine, like the waves trailing behind a boat. At a recent American Geophysical Union meeting in San Francisco, her team put forward the initial results from its first large set of measurements, which the researchers are still analyzing.

The lidar studies observe that modern turbines typically rest on tall towers about 250 feet (80 meters) high, and the strongest part of the wake stretches downwind from a turbine about two or three times the height of the tower. The swirling turbulence from the turbine spreads out as it travels downwind, the lidar study found. "We're still trying to figure out where, and in which circumstances, the wake hits the ground," Lundquist said.

- **A Need for Answers:** While it has many positives, the turbines' influence impact on temperature could pose a threat to crops. A rise in the temperature at night could increase the amount the plants respire—a kind of exhalation at night time when plants release some of the carbon they took up from the air during the day. That would ultimately lead to the plants picking up less carbon, as a consequence losing out on the benefits from their growth during the daytime. It could be seen as a mishap.

"The possibility underscores the importance of the research for the agricultural community. Farmers for years have leased portions of their fields for wind turbines as a way to boost income, but they have wondered what effect it might have on crops," Lundquist said. "It's a big choice for farmers to make," she said, so they hope to start getting answers soon.

Another critical study from 2010, led by Somnath Baidya Roy, an atmospheric scientist from the University of Illinois, studies the weather around a wind farm and affirms the aforesaid situation. His study found that in southern California, night time temperatures were higher downwind from wind turbines.

Roy agreed that the effects on crops will be complex, and "one good thing could offset another bad thing."I think the frost protection effect for crops is going to be a really good thing," Roy said, adding that it might outweigh other effects.

Turbines have a variety of effects on plants and different plants require different growing conditions. Because of this very reason, Lundquist throws a word of caution that what may help Iowa corn might not help other crops in other places. "Wind energy offers us great potential for renewable energy," she said, "we just have to be clever and sensitive about how we deploy it.

Another ground-breaking research carried out by the TNAU Agritech portal, states that wind direction and velocity have a significant influence on crop growth.

The study carried out states the following beneficial impact of wind:

- Wind increases the turbulence in the atmosphere, thereby increasing the supply of carbon dioxide to the plants, resulting in higher success rates of the process of photosynthesis.
- Wind alters the balance of hormones.
- Wind increases the ethylene production in barley and rice.
- Wind decreases gibberellic acid content of roots and shoots in rice.
- Nitrogen concentration in both barley and rice increases with an increase in wind.

Wind influences crop production in two ways:

- Physiological impact; and
- Mechanical impact.

6.4.1. Physiological Impact

- Increases transpiration especially cuticular transpiration than stomatal transpiration.
- Hot wind accelerates the drying of plants by replacing humid air by dry air in the intercellular spaces. For example, rice crops during June–July months shows tip drying.
- Wind increase turbulence in the atmosphere and the availability of carbon dioxide. It results in increased photosynthesis.
- Beyond a certain wind speed, the rate of photosynthesis becomes constant.

6.4.2. Mechanical Impact

- Strong wind damages the shoots;
- Lodging due to strong wind (paddy, sugarcane, banana, etc.);
- Flower and fruit shedding;
- Crops and trees with shallow roots are uprooted;

- Cold wind causes chilling injuries;
- Causes soil erosion;
- Soil deposition causes poor aeration in the root zone.

6.5. ADVANTAGES OF WIND POWER AS A RENEWABLE ENERGY

Wind energy is one of the fastest-spreading energy sources in the world. It is one of the most easily adaptable energy forms and offers a variety of advantages (Figure 6.8).

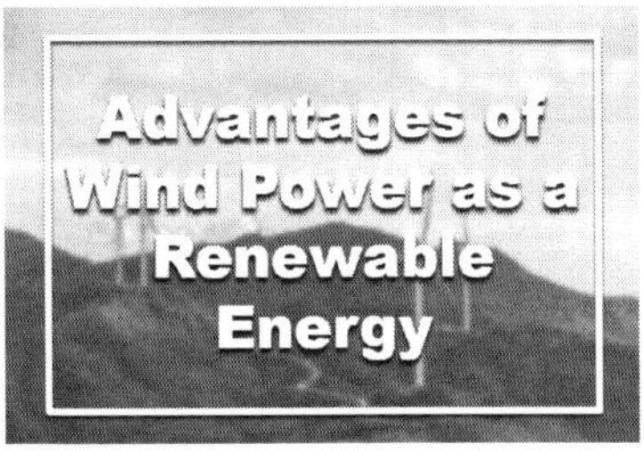

Figure 6.8: Advantages of wind power as a renewable energy.

Continuous efforts are underway in ground-breaking research with the aim of addressing the challenges to its greater and widespread use. The key points which highlight the benefits wind energy provides are discussed in subsections.

6.5.1. Advantages of Wind Power

Wind power is economical and comparatively low-priced than other energy resources. The land-based utility-scale wind is one of the lowest-priced energy sources available today, costing a mere 1–2 cents per kilowatt-hour after the production tax credit.

Wind energy alleviates the projections of price uncertainty that fuel costs bring to the conventional and non-renewable sources of energy. The electricity from wind farms is sold at a fixed price over a long period (e.g., 20 years or more), and its fuel is readily available free of cost. Thus, it turns out to be a low-cost and pocket friendly energy source.

The Wind energy sector has created a plethora of employment opportunities worldwide and the U.S. wind sector alone employs more than 100,000 workers, and wind turbine technician is one of the fastest-growing

American jobs. According to the Wind Vision Report, the wind has the potential to support more than 600,000 jobs in manufacturing, installation, maintenance, and supporting services by 2050.

Wind enables countries like the U.S. to boost industry growth and their competitiveness in the global market. New wind projects account for annual investments of over 10 billion dollars in the U.S. economy. The United States has a vast domestic resource and a highly-skilled workforce and can compete globally in the clean energy economy.

It is a clean fuel source. Wind energy does not cause any pollution in the air like power plants that heavily dependent on the combustion of fossil fuels, such as coal or natural gas, which emit particulate matter, nitrogen oxides, and Sulphur dioxide—causing human health hazards and also pose grave economic and natural damages. Wind turbines do not produce atmospheric emissions that are responsible for acid rain, smog, or greenhouse gases.

The wind is a domestic source of energy. Every nation's wind supply is abundant and inexhaustible. In the course of the past 10 years, the U.S. wind power capacity has grown at an impressive 15% per year, making wind the largest source of renewable power in the United States today.

It is sustainable. The wind is actually a form of solar energy. Winds are caused by the heating of the atmosphere by the sun, the rotation of the earth, and the earth's surface irregularities. For as long as the sun shines and the wind blows, the energy produced can be harnessed to send power across the grid.

Wind turbines can be built on existing farms or ranches. They do not need any special land acquisition, and this greatly benefits the economy in rural areas, where most of the best wind sites are found. Farmers and ranchers can continue to work the land because the wind turbines use only a fraction of the land. Wind power plant owners make rent payments to the farmer or rancher for the use of the land, providing landowners with additional income.

6.6. CHALLENGES OF WIND POWER

Like every other energy resource, wind energy to comes with a set of drawbacks. The following points will discuss the same.

Wind power must still compete with conventional generation sources on a cost basis. Even though the cost of wind power has decreased dramatically in the past several decades, wind projects must be able to compete economically with the lowest-cost source of electricity, and some locations may not be windy enough to be cost-competitive.

Sites with good wind are often found in remote areas or on the outskirts far from cities, where the generation of electricity is needed. Transmissions lines must be built to bring the electricity from the wind farm to the city. However, building just a few already-proposed transmissions lines could significantly reduce the costs of expanding wind energy.

Wind resource development might not be the most profitable use of the land. Land suitable for wind-turbine installation must compete with alternative uses for the land, which might be more highly valued than electricity generation.

Turbines do cause a low level of noise and esthetic pollution. Although wind power plants have relatively little impact on the environment compared to conventional power plants, yet concerns loom large over the noise produced by the turbine blades and visual impacts on the landscape.

Wind turbines can impact local wildlife. Quite often birds have been killed by flying into spinning turbine blades. While with new developments most of these problems have been resolved or greatly reduced through technological advancements concerning the same as also by careful positioning of Wind Turbines.

Bats, too, have been killed by the rotating turbine blades, and researchers are trying to find solutions to eliminate such impact on the species. Like all energy sources, wind projects can alter the habitat on which they are built, which may alter the suitability of that habitat for certain species.

6.7. CONCLUSION

The future holds bright prospects for all forms of renewable energy due to the heavy consumption and near depletion of fossil fuels and all forms of renewable energy. The primary objective of employing wind energy in agriculture is to rapidly advance economic development, maintain energy security for future generations, improve, and spread easy access to energy, and negate and combat climate changes brought about by centuries of conventional energy methods and practices.

Strong support by governments all over the world, and the increasingly favorable economic conditions have made possible the use of sustainable energy sources like wind energy to ensure economical, reliable, and sustainable adoption of modern, clean energy sources like wind energy.

Significant achievements like the generation of electricity and, ease of farming have been realized by deploying wind energy in agricultural practices.

However, various obstacles and challenges have also been identified in the adoption and development of wind energy. Policymakers and innovators must step forward and help this clean energy source to be the future of sustainable and organic agriculture. The world is faced with a choice of energy futures with biomass, thermal, and solar energy giving stiff competition to wind energy.

Wind energy today is being used in a variety of applications on farms and ranches the world over, and there are plenty of opportunities to expand its use in the future. In-depth research, development, demonstration, deployment, and aggressive commercialization would help build a continuum flowing in the field of wind energy.

REFERENCES

1. Alireza, K., & Omer, C. O., (n.d.). *Energy Harvesting: Solar, Wind, and Ocean Energy Conversion Systems (Energy, Power Electronics, and Machines)-PDF Drive*. [Online] Pdfdrive.com. Available at: https://www.pdfdrive.com/energy-harvesting-solar-wind-and-ocean-energy-conversion-systems-energy-power-electronics-and-machines-e185161677.html (accessed on 5 May 2021).
2. Ashara, N., (2016). *Sustainable Energy Applications in Agriculture*. [Online] Slideshare.net. Available at: https://www.slideshare.net/zarandean/sustainable-energy-applications-in-agriculture (accessed on 5 May 2021).
3. C-DAC, (2021). *Wind Energy for Water Pumping*. [Online] Vikaspedia.in. Available at: https://vikaspedia.in/energy/energy-production/wind-energy/wind-energy-for-water-pumping (accessed on 5 May 2021).
4. Energypedia, (2020). *Wind Energy in Powering Agriculture-Energypedia.info*. [Online] Energypedia.info. Available at: https://energypedia.info/wiki/Wind_Energy_in_Powering_Agriculture (accessed on 5 May 2021).
5. Michele, (2019). *Fundamentals of Wind Turbines | Wind Systems Magazine*. [Online] Windsystemsmag.com. Available at: https://www.windsystemsmag.com/fundamentals-of-wind-turbines/ (accessed on 5 May 2021).
6. Svickova, P., (2020). *Advantages and Disadvantages of Wind Power—Barani Design Technologies*. [Online] Barani Design Technologies. Available at: https://www.baranidesign.com/faq-articles/2020/2/20/advantages-and-disadvantages-of-wind-power (accessed on 5 May 2021).
7. Wikipedia, (2021). *Wind Power-Wikipedia*. [Online] En.wikipedia.org. Available at: https://en.wikipedia.org/wiki/Wind_power#History (accessed on 5 May 2021).

Chapter

7

Innovations and Challenges in Agriculture

CONTENTS

In this chapter, the innovations and challenges in agriculture have been discussed in detail. The various challenges to agriculture like small and fragmented land holdings, the challenge of non-availability of good quality seeds to poor farmers, problem related to agriculture, lack of modern machinery use by farmers and soil erosion due to climate change, not enough storage facilities and scarcity of capital available to farmers at the right time when they need the most, etc.

From the consumer point of view, there are food safety concerns related to genetically modified (GM) seeds, and challenges related to selling that produce and exporting to other countries have been discussed in detail.

In this chapter, there is also a detailed discussion about innovations of various technologies and devices that have and there is a chance that they can completely transform agriculture has been discussed in detail.

In this chapter, there is a discussion on some innovations like agriculture by the use of aircraft, using the sun to generate freshwater from the sea and using it for agriculture, vertical farming, use of drones, bee vectoring and many other latest innovative technologies has been done.

7.1. INTRODUCTION

There has been a long history of agricultural research and advancement which has managed to improve agriculture and support the feeding of our increasing population around the world. This, together with the rapid development of food technology has caused more people than ever before to be supplied with food, in volume and quality.

The food processing of today is considered harmful for its environmental impact and socioeconomic burden that includes emerging health problems like diabetes and obesity at the same time. It is increasingly not acceptable that the number of people who are under the tragic and horrible conditions of starvation is rising while obesity is rising significantly across the world (Figure 7.1).

Figure 7.1: Innovations and challenges in agriculture.

The world has reached a pivotal moment in our food systems. Each new foodstuff must be created for the needs and the society of the whole population that will follow tomorrow. The people must change our attitude towards food production that allows for these current issues, and there are no other solutions, producing fewer resources and doing more to protect the environment, can also benefit human beings.

Innovation in agriculture will lead to a reliable and equal return over the long term to the countries around the world. In reality, over the last decade, there have been new types of innovations that put together mission and benefit, as expressed in the increasing number of businesses creating their policies for corporate social responsibility and the creation of SMEs (small and medium-sized enterprises) with a clear social and environmental emphasis.

The volatile prices of oil, low growth, environmental issues, and growing customer demand are some of the dynamic obstacles confronted today by the crops industry, as well as decreased profit margins for farmers. To meet the needs of Europe and the world, sustainable crop solutions are needed for agricultural innovation, and so that food production can be developed economically and socially and environmentally, food production must also need to be done more heavily in Europe.

As part of their efforts to improve crop production and sustainability and yield as well as food competition, future researchers would have to think of more effective land-use practices prior to and post-harvest management practices and create technologies that conserve natural resources while remaining accessible.

Innovative approaches, more or better ways of pest management methods are used routinely now. Such practices are generally obligatory, but they are more prevalent and practiced in some countries, such as integrated pest management (IPM).

To broaden the scope of our applications, there is a need to have the capability and capacity to seek out new solutions. In addition to production challenges, farmers have to deal with problems of environmental sustainability.

While there are many other occupations that are as difficult as farming, only a few of which may be so rare that they can prevent someone from being successful, farming is described as difficult. To ensure their existence for generations, farmers have been at the forefront of an ongoing search for different ways to overcome a wide range of ecological threats. And because of this, they have an incredible degree of variance in personality and attitudes, over the years, they have become very malleable people.

Despite the rise of access to information, however, a myriad of different problems like age, scarcity of skills, previous agricultural experience, access to resources, no knowledge of the latest techniques, age-old issues shortcomings for modern farmers remain constant. And because of these emerging threats related to the environment and the world economy, there has been much more pressure on older workers to retire.

7.2. CHALLENGES IN AGRICULTURE

7.2.1. Small and Fragmented Land Holdings

Land reform, in general, has played a prominent role in the discussion of resolving issues of economic and social problems in rural areas, especially rural poverty in developing countries. Land redistribution may have the advantage of providing land for poor citizens, isolated villages with an extra piece of property to expand the size of their farm plots but could contribute to land fragmentation.

If the distribution of landholdings were expanded, the whole range of properties would become increasingly uneven. Fragmentation of the field may also reduce the supply of optimal factors, making land production lower yielding than it would be. The following may play a role in this could include decreased efficiency due to increased travel time, useless boundary areas, unacceptably small measurement areas, and the fact that one cannot use the equipment like harvesters.

The weak agricultural production in India can be related to the fact that the land is over-divided or poor land management, where a small tract of land is owned by small, marginal farmers.

While India has significantly increased its yields by 3 fourth in sugar cane yields, 57 th in cotton, 118 th in pulses, and 51 st in rice, India is the leading aggregate producer of each such crop. (Economic Survey 1998–99.)

The main factor influencing costs in India is the amount of water added to the farmers placed into each crop, which is more than is used by farmers in Egypt to grow the same amount of cotton. For example, they are using approximately 25 times more water for irrigation to grow cotton.

In reaction to land fragmentation being considered to be a source of possible detrimental economic and social impacts, the Finance Minister in 2000 allotted the 5 million rupees over a 5-years to consolidate existing plots of land into one large holding. It has been suggested that in several sources that the Planning Commission of India has been almost unsuccessful in the same respects. (Economic Survey 1998–1999) (Figure 7.2).

Figure 7.2: Small and fragmented landholdings.

The vast amount of uncultivated net area in India is segmented into small and impoverished land holdings that do not make sense to use it economically. There is a reason for the present way the agricultural inheritance laws are primarily responsible for the fragmentation of land. The land is equally divided among the children of the father, who returns to the family to start their farm.

Although the allocation of land does not shift in one huge group, the fact is that it is highly dispersed. These properties can vary, and they influence what kind of crop can be grown in the region: fertility, distribution of nutrients, texture, and countless other factors in the environment.

When all the land has been shared evenly among two sons, each of them receives a smaller portion of it. In this way, the accumulation of capital happens, which allows the holdings to become less consolidated and more diffuse with each subsequent generation. This will lead to decreased productivity of the farm because the holdings are in poor condition.

The problem that often emerges from poor land-use fragmentation is the poor distribution of water supply. It is an impossible feat to combine all the scattered parcels into a single holding. Additionally, the most fertile land is being used in the process of setting borders.

The only way this issue of Indian agriculture can be solved is by the implementation of consolidation, which means consolidating the scattered land and creating more large farms. This was a detriment since they did not see much success. There are a few additional solutions to this issue, and one of these is to create cooperative farms in which farmers combine their money and distribute profits.

These troubling situations have arisen because policies do not permit the preservation of family wealth to go into the next generation. To ensure proper family unity, the land in an equal sum is divided between all male members. This dispersion of land does not involve aggregation or consolidation, but instead, after being dispersed, favors slow growth.

Since fertility levels are high in some areas and low in others, those areas should be allocated in differing quantities to be efficient. This means that where there are four tracts of land to be divided between two sons, the smaller tracts would be assigned to the two of each of them. In this way, the accumulation of capital happens, which allows the holdings to become less consolidated and more diffuse with each subsequent generation.

Many farmers split their landholdings by subdividing and fragmenting the possessions are among many separate owners, thereby decreasing production and adding to our country's backwardness. Agricultural, animal, livestock, time, and labor are frequently squandered in transporting crops, implements, horses, and livestock between parts of farmland.

If there are several small and separate fields to be irrigated, so it would be impossible to do so. In addition, a large amount of cultivated ground is consumed by such uses as barriers or hedges. In such conditions, the farmer would not have the time to focus on fine-tuning his or her operation.

The only solution to this vexing dilemma is to merge holdings, which involves reallocating dispersed holdings and creating farms with only one or a few plots instead of the multitude of patches formerly in each peasant's hands.

To deal with this issue, there are various solutions such as group farming in which farmers combine their resources to ensure financial benefit, or in which farmers increase their profits through group initiatives.

7.2.2. Lack of Mechanization

Despite the large-scale mechanization in some areas, much of agriculture is still carried out by hand, using simple and traditional tools like wooden plows, sickles, etc.

They use no mechanical machinery for plowing, sowing, harvesting, threshing, and mowing; thinning and pruning are done by hand, and threshing and transporting the crops are done by machine. This is particularly true for small and marginal farmers and poor people. Willingness to work expenditure of people is borne at the cost of tremendous losses in production and meager earnings per person.

Widespread loss of human labor should be avoided in farming to make farming more practical, and mechanical advances must be implemented to assist the work. Agricultural machinery and equipment are essential inputs for effective and timely crop management, allowing more than one crop to be planted in each plot and consequently growing yield.

After Independence, little progress was made in the development of Indian agriculture by means of technology. It was quite a challenge for mechanization to come along with the Green Revolution in the 1960s. Over the years, farmers have used less effective and conventional farming tools to supply higher levels of productivity. A large portion of their investment has been directed to replacing these with newer and more productive alternatives, such as tractors, harvesters, tillers, and mowers.

An abundant supply of raw materials has enabled the development of thriving manufacturing industry for agricultural machines.

This growth was due to the increased use of rotary, mower, harrowing, and self-type grain drills as well as part of machinery such as motors coupled with irrigation pumps.

It is now in the industry's interest to urge farmers to use equipment of a technologically advanced design as rapidly as possible while conducting operations in a precise and timely manner to increase output and save costs (Figure 7.3).

Figure 7.3: Manual labor working in agricultural fields.

Sources: Image by Piqsels.

7.2.3. Lack of Agricultural Marketing

Agricultural marketing remains a concern in India, even though it has been significantly developed over the last several decades. In the absence of viable marketing facilities, farmers have to sell their products in the local and regional markets, where middlemen and traders absorb their low-cost and the farmers receive an even lower price.

Most of these farmers find themselves in situations where they must sell off their goods at distressed prices due to socio-economic conditions. Nearly all of the country's small-town farmers are normally financially involved with moneylenders, and these usually buy their crops with the cash they have to return.

There is nothing left for the poor farmer to do but sell his produce to fulfill his financial obligations and pay his debt, so he will sell it to whoever will pay him what he thinks is fair. A large part of the results concluded that the manufacturers struggle to sell their goods at favorable prices and that much of the terms on which they do are unfair.

Masters and middlemen have taken over the selling and trading of agricultural goods in the absence of an organization system in which only private traders and certain individuals can participate. This raises the price that the buyer must pay because the new regulations to compensate the middlemen fairly for their services, but not the people who are providing the goods.

The market studies and other market analysts indicate that middlemen can take as much as 48% of the money paid for rice, as well as 52% of the money received for groundnuts, and can be as much as 60% of the money obtained from selling potatoes to consumers.

In a way to prevent the exploitation of farmers by money lenders and middlemen, the government has implemented controlled markets. Generally, these markets provide a process of competitive bidding that prevents various unfair business practices from taking place, provides producers with fair rates, ensures profitable participation and proper compensation for participants, and honest pricing of products, and appropriate settlement, and controlled scale of grievances, and assists in the evolution of industrial self-wide measurements.

7.2.4. Inadequate Storage Facilities

In the village areas of developing countries, there are typically no food storage facilities, and they are both extremely deficient. Since this puts such strain on farmers, they must either sell their goods after they have gotten their yields or be forced to sell at the price lower than market prices. The farmers are robbed of their actual income by such as a result of this distress selling. The implementation of scientific storage is especially important to ensure both the protection and the continued use of agricultural products for the producers and the consumers.

In the current situation, there are a variety of different organizations involved in warehousing and storage programs. One of the principal tasks that the Food Corporation of India (FCI), the Central Warehousing Corporation (CWC) and the State Warehousing Corporation (SWC) are involved in is the expansion of storage and distribution. These organizations work to maintain and stockpile buffer supplies, which can be used in the hour of need. The program for setting up the Grid of Rural warehouse, which started in the year 1979–1980, is now being executed by the National government.

This system is useful for farmers with proximity to their farmland, and for those in specific need of storage, such as the small and marginal producers The Working group on rural storage facilities recommended a plan that the establishment of a network of Rural Storage Centers to meet the needs of farming and rural industry.

7.2.5. Lack of Availability of Good Quality Seeds

In addition to seed, additional inputs, such as fertilizer, land improvements, are important in growing yields and increasing agricultural output. In terms of having quality seeds, the supply of quality seeds is just as important as much as the manufacturing of high-quality seeds. But high-quality seeds are out of scope for farmers because they are too expensive for those who do not have large, specialized farms because of the existence of enormous seed and chemical seed expenses.

However, Indian agriculture will show even more of the future development as the country expands its agricultural base. Indian agriculture is likely to offer the seeds of increased potential as the country increases its agricultural base. It is not only the task of the seed industry to ensure that the availability of enough good seed and to also attain diversification of varieties to meet different agro-climatic zone requirements.

The food and nutritional objectives of the government program for farmers are structured to assist seed suppliers to provide ample quantities of high-quality seed to be available at the required time and at market rates that are both usable and equal.

The expansion of agriculture was primarily achieved with the use of limited-generation practices of seed multiplication. This method is capable of distinguishing three types of generations, namely breeder, licensed seed, and base. It is a simple seed and a starting seed used to produce the first seed as well as the subsequent seeds. As in the example above, foundation seed breeders produce seed, and as well as creating it, they keep track of and tracking seed is critical in plant breeding.

For Indian farmers, the most difficult problems are consistency, price, and the provision of seeds on time are things that often hold them back. The majority of Indian farmers are poor and thus are economically unable to buy seeds because they must spend heavily on pesticides and fertilizers. This is because they are mostly based on conventionally-seeded seeds.

Other types of seeds do not need frequent purchasing, so farmers do not need to do not have to buy them annually. These seeds are therefore non-poisonous, and therefore beneficial to the ecosystem. They are safe and have high nutrition levels of natural ingredients. However, the biggest issue with these seeds is that they have a poor yield, which hinders productivity. Although it is true that indigenous seeds have lower production relative to other varieties, indigenous seed strains are unable to keep up with the country's growing population size.

This can be achieved in developing countries, where there is an already severe food deficit, where conventional seeds can be combined to better meet the need for the malnutrition crisis. The faster-growing and more efficient hybrids are better for the production process.

The seeds of these plants are less likely to be infested by rodents and disease than those from traditional seeds. Another good feature of hybrid variety is that they are more capable of accommodating changes in the environment. There are still some disadvantages of hybrid seeds. Hybrid seeds do not have an off-sown yield; therefore, farmers must buy each year's supply rather than replenishing it, and this downside could be harmful to an off-farm source that cannot be used to offset it.

If the seeds have been grown into plants, they cannot be saved, because they have a feature that stops them from germination, so the farmers have to rely on the large organizations as seeds can no longer be used for next year's planting. This elevates their monetary expenditures.

The government has taken measures to deal with these problems with genetically modified (GM) crops. GM seeds could provide substantial relief to hunger in India, as they have the potential to increase agricultural productivity significantly.

In all of this, it is more environmentally friendly, it uses less fertilizer and pesticide than conventional and hybrid crops. A significant disadvantage of GM seeds is that they may disturb the natural balance of traits in the ecosystem.

7.2.6. Soil Erosion

The scientist has been alarmed by soil erosion because of its negative effects on the nation's food security. While also, it is a serious problem for agriculture. Soil is the most valuable resource because of all, and that leads to the high yields of all crops. Soil erosion also represents substantial damage to the fertility of the soil.

There are many ways in which soil erosion can be carried out, such as water, livestock, etc. Water is the most important, and since it carries away 90% of the sediment from the land and guides the remaining 10% of the erosive phase. Forgetting to preserve away the topmost layer of soil makes a farmer leave behind residue, which depletes the soil fertility. It is more likely to occur due to overgrazing, wetland/land growth, building small dams, and changing cropping practices, but all of these could be avoided with suitable measures such as reforestation, check-overgrazing, terracing, and contouring.

A lot of lands in regions where the soil is vulnerable to erosion is found are over-exposed to both by wind and water. There is a need to regularly feed and care for this region to keep it healthy and fertile (Figure 7.4).

Figure 7.4: Soil erosion due to heavy rain.

Sources: Image by Flickr.

7.2.7. Inadequate Transport Facilities

As one of the major problems with Indian agriculture, transportation is prohibitively costly and unreliable, many agricultural activities are doomed to stay isolated. Even today, there are lakhs of villages that have to rely on difficult to reach or absent roads to have market places that do not pass through them.

Most of the rural roads in Kutcha (soil or mud paths) are inaccessible in the rainy season. This causes hardships for the farmers who are unable to take their produce to the broad city market, forcing them to be pushed to the more local rural area where they must sell it at a depressed price, and disadvantage for the city inhabitants. There is an enormous amount of money required to do this job, and it will take a large amount of money to connect each village with paved roads.

7.2.8. Lack of Irrigation Facilities

Despite becoming the world's second-largest irrigated country after China, irrigation covers just one-third of India's cropped land. Unreliable rainfall is a significant challenge in a country like India, where the primary irrigation comes from intermittent and unpredictable irrigation sources, such as canals, reservoirs, and the likelihood of more challenges from secondary run-off and groundwater is great. India will not realize any long-or even the country's long-term economic progress in agriculture unless the current irrigation coverage of half of the cropland is successful.

Canals are ideal for the cultivation of many kinds of plants, but to safeguard against issues like salinity, however, care must be taken in irrigation, particularly in regions where the canal water is used (Figure 7.5).

Figure 7.5: Picture depicting irrigation of crops.

Sources: Image by Pixabay.

7.2.9. Manures, Fertilizers, and Biocides

Despite the fact that Indian lands have been farmed for thousands of years, no effort has been put into their restoration and replenishment. A slight but persistent decline in soil depletion has caused soil erosion and depletion of the productivity of crops.

The world's crop yields are considered very weak relative to those of virtually all other nations. It can be dealt with by the introduction of more manure and fertilizer as it is a serious problem.

People are much like soils. In the field, manures, and fertilizers have the same function: manure and fertilizer serve to the soil and the same manner as food serves in people. What this sentence has shown, though, is that just as a safe and groomed body can do any task, balanced, and well-nourished soil can deliver the best yields a nutritious and great harvest.

If one uses the claim of expanded fertilizer utilization being responsible for 70% of agricultural output, it has been calculated that up to two-thirds of global agricultural output development is due to its existing use.

In this way, an increase in the use of fertilizers is an indicator of agriculture's financial and ecological well-being. Thus, there are considerable logistical problems when it comes to delivering manure and fertilizer to major population centers like India, which are riddled with peasants who do not have the means to procure these inputs. The manure contains the largest concentration of nitrogen, phosphorous, and potassium in the soils.

Since much of it is used as cooking fuel, it is in the form of dung cakes; this could be a little less common than regular cow manure. This new population growth and increased demand for firewood have exacerbated the crisis. Chemical fertilizers are typically cost too much for small farmers and maybe beyond their means, so they are seldom used. When it comes to fertilizers, this is a serious and complicated challenge.

It is believed that organic manures are essential for maintaining the health of the soil. Most of this waste material can be recycled in the world has 650 million metric tons of rural and 160 lakh metric tons of urban compost and is currently being neglected. Expanding on the future solution to this would remove pollution and fertilize the soil to address the issue of waste.

The government has provided strong financial incentives for using chemical fertilizers, providing these benefits particularly in the form of generous tax credits and bounties. Between the time of the independence and the separation of South and North America, there was little or no use of chemical fertilizers in South America. The agricultural department introduced more sustainable farming practices and this, such as purchasing manure, and other alternatives, fertilizer use went up substantially.

The pests, germs, and weeds caused one-third of the overall produce loss at the time of independence, which was offset by planting and fertility enhancers. Biocides (pesticides, herbicides, and weedkillers) are employed to protect fertility and reduce the lack of yield. A large percentage of these food crops have been spared from being unnecessarily wasted thanks to the improved use of these inputs. The indiscriminate use of biocides has caused a wide-ranging and pernicious environmental degradation but has also exacted a toll on the environment.

7.2.10. Scarcity of Capital

Agriculture is as well as other industries must have funds; it is essential for society's growth. The increasing importance of capital input in farming, due to the increased use of new technologies, is pushing up the price of agricultural products. He is under obligation to raise the level of his stock and land and/animal assets, so he therefore must borrow money to support the development of additional animal and crop growth.

Also, the key providers of capital to the farmer are the moneylenders, brokers, and agents who lend him money and buy low-in terms of interest and get crops at low prices.

The effect of providing loans by government-backed banks can be seen in the fact that the moneylender has decreased their position in agricultural lending but is still the major financial provider of agricultural loan financing. A major expansion has resulted in the availability of credit facilities in rural areas with the provision of both public and private financial institutions, national and industrial establishments including central banks, as well as co-ops, state, and federal cooperatives, have strengthened their capacity to offer loans to farmers.

7.2.11. GM and Food Safety Concerns Among Consumers

The customers of both nations are more concerned about the methods of food production and processing, and their food quality, which is why concerns about GM foods have grown in China and the USA. Regardless of the positive or negative impacts, these methods may, or possible impacts may have, a majority of the food and fiber currently produced around the developed world is produced using genetic engineering, which will continue to increase in abundance over time (Wong and Chan, 2016; Raman, 2017).

To address that issue, 2,063 respondents from all over China's provinces, with bi-national national backgrounds were interviewed and it was discovered that the majority of those who supported, opposed, or remained neutral to, to these types of products, respectively, were 11.9, 41.4, and 46.7% (Cui and Shoemaker, 2018).

These older, more economically well-off, and highly educated respondents were the most opposed to the use and consumption of foods such as pizza, shrimp, lima beans, flounder, and potatoes (Cui and Shoemaker, 2018). According to the current findings, results, Chinese, and American consumers do not differ significantly in their behavior. According to a new poll on the public's views of GM food, the number of Americans who voiced concern about the dangers has remained largely stable, although confidence in the benefits has dropped.

The 59% of respondents out of 2539 people answered that GM foods were likely or highly likely to pose health risks for the US population in 2018. Also, over time, this belief increased from 39% to 49% and was substantiated by 10% more recent data, albeit data which demonstrated that genetic engineering "can cause long-term problems with human health" (Funk et al., 2018).

While both China and American and international resistance to GM grains and fruits and vegetables are significant challenges, the trend itself presents a significant difficulty to Chinese and American farmers and consumers (Wong and Chan, 2016).

In both countries, the farmers both countries are caught between, falling profit margins are linked to cheaper imports, and market aversion to genetic modification (Gerasimova, 2016; Butkowski et al., 2017). Instead of focusing on numerous ways to potential solutions, the world's farmers often only take genetic modification as an option and assume that every other potential alternative has already been exhausted.

And, because of herbicide-resistant weeds and pesticide-resistant pests, most agronomists and agricultural economists claim that commonly accepted genetic modification technologies lower prices, decrease the usage of traditional fungicides, herbicides, and pesticides, and, in some cases, provide the only agronomic choice (Gerasimova, 2016; Wong and Chan, 2016; Giannakas, 2018).

Even though new research reveals consumers are more accepting of GM crops and information has thus led to a drop in resistance to their use and consumers are provided with important information in regard to the healthy use of GM crops as well, GMOs have still been highly controversial for years and opposed due to potential health risks (Butkowski et al., 2017). There are no government-issued GM rice varieties in existence as of any developed, as of the variety as of the year in 2018.

This could mean that no GM rice is for sale to the public or export can be made available. GM rice is most often sold locally, but some of it is also exported to other countries as part of large crop production for both commercial and export markets.

The European Union expanded research on Chinese rice after discovering contamination of GM rice in volume shipments (25 cases in 2013 alone), and subsequently banned the selling of Chinese-sourced rice in Europe, stating environmental safety issues (Cui and Shoemaker, 2018).

In the case of the food additive worries, American, and Chinese consumers both want to know how the additives they consume in small amounts daily doses will affect their health in the long term. I n the foods the Americans consume over a day, while the other half (49%) feels that the current additives do not cause much harm (though it may take years for these beliefs to manifest themselves into serious health problems as per 2018 Pew survey (Institute of Medicine, National Research Council, 1998).

A plurality of those polled believed that food contaminated with antibiotics and hormones (80%), pesticides (79%), artificial preservatives (78%), and artificial coloring (69%) presented some danger (Funk et al., 2018). Concerns about pesticide contamination, soil, and water contaminants, toxins, antibiotics, and hormones were also raised in surveys of Chinese consumers who were directly asked about anxiety linked to food additives (Veeck et al., 2015; Si et al., 2018).

Due to the nation's rapid speed of food system growth and a large number of small food producers, processors, and distributors that populate the system, China has faced significant challenges with food safety (Chung and Wong, 2013; Shao, 2013).

The problem of Fake food could contribute to the population's current food risk-reduction anxiety, which in turn may hinder coping mechanisms for food safety, as it could make people less confident of the quality of their food. It is a waste of money to pay to trust a brand name that they are authentic because customers cannot be certain of the claims that are made about the product.

In the same way, stickers like environmentally friendly, locally manufactured, or organic and handled as far away from the source as possible, because the consumer cannot be sure about the sustainability of such words if the products cannot be obtained from a place which is in keeping with environmental ideals (Veeck et al., 2010).

While achieving international food safety regulations and standards has always been difficult, this difficulty has risen in recent years, as it becomes more important with increasing problems with foodborne illness (Fu, 2016).

The United States announced plans to intensify efforts to control food risks in 2019, as did China.

To employ modern, more effective technologies to avoid and track contamination in food, the Food and Drug Administration (FDA) launched an initiative called a New Era of Smarter Food Safety (FDA, 2019). In the background of a communication campaign called Stand True to Our Founding Mission to improve the Food Safety Law of 2015, China declared a crackdown on food safety (Wang, 2019).

Using recent efforts to fight against counterfeit and contaminated food, the country food products has identified nearly 70,000 violations, which have resulted in the state and $130 million in fines for companies in the recent 12 months of the equivalent of almost a year of money being collected from the penalty imposed for these violations over the two-month period of June and December of this year (Wang, 2019).

Food threats can hardly be mitigated altogether. Steps should nevertheless be taken to boost trust in food systems. New technologies, including blockchain IDs, are useful in helping to find and remedy food safety issues and smart sensors, artificial intelligence (AI), and the internet of things (IoT) are being developed in hopes of assisting with food safety protocols (FDA, 2019).

Both China and the United States should do everything possible to develop and implement measures to increase the traceability of food products and make the food supply chain more transparent. Traditionally, there have been close links between our agricultural systems, the levels of the US and China, and improving food safety is a shared concern between us and them (Zhang et al., 2008).

The technological advancements of the two countries could enhance food safety worldwide as their goods are exported to hundreds of other nations. If the agricultural sectors are to continue to evolve, retailers must feel secure in their business.

7.3. SOME OF THE INNOVATIONS IN AGRICULTURE

These estimates are that there are 570 million farms in the world, but agricultural and technology and the means of the financing were involved in each other in a complex, multifaceted vortex in 2014, the prediction is that technical funding generated around $570 million as a result of all the countless farms.

Although a significant investment has been made to help entrepreneurs and people working in uncharted markets like agricultural technology and the realm of environmentally sensitive agriculture and even in general, some goals like increased productivity in land and water use have not changed over the years.

7.3.1. Data Preserved in Soil

With regards to conventional farming models, perhaps the main factor in deciding supply potential is the capability of the volume and nature of the land. Knowledge about the future assumptions, expectations, thus, must be assembled from current information and followed by additional data that details out the same area.

More recent and cutting-edge methods, such as automated soil mapping in the impoverished locations of Africa, are critical for advancing agricultural intensification and for those who want to pursue natural resource management. In Google Earth, the public can explore these virtual maps, so these map details are viewable by the general public.

7.3.2. Innovative Agriculture Moving Underground

Due to import dependency, virtually all fruit and vegetables are vital to the running of Singapore. It would delight consumers to find out that their bread, sautéed in a limited amount of fat, served with tender baby spinach and radish, is freshly cooked and locally grown. In addition, not only has it been established in a TV maker's first legal farm, but it has also flourished in a big corporation best known for electronics that is Panasonic.

While they may have developed a maximum annual output potential of 3.6 tons of farming tech, not only are the early Panasonic brands cultivating practices rather than profit-driven ones but there are other, more high-tech farming solutions.

The growth of strawberries is occurring in Dubai, while in other parts of Japan, Toshiba, Sony, and Fujitsu are using the former clean-room facilities in semiconductor factories for the cultivation of lettuce. Because of the unique hydroponic LED illumination, these no-treated greens are grown and harvested twice as fast as regular field greens for greater yields.

7.3.3. Greens Fed on Rainbow Waste

In the field of agriculture, reciprocal efforts (corresponding to mutual cooperation) are gaining prominence. This term Hydroponics describes farming methods that use water to nurture crops without the use of soil and fertilizers. Fisheries and agriculture are not only mixed in aquaculture but are also fully incorporated in such a recirculating method.

The Bio-ecosystem at Blackford Farm in Somerset-Europe's first fully developed aquaponic farm entirely grows vegetables without using chemicals or pesticides, but with the aid of bees and worms.

The fish are responsible for around half of the plant-based nutrition, due to wastewater generated by the aquaculture of waste fish metabolites are firstly broken down into nitrogen and then reused by way of basic biological therapy, with the nitrogen-containing solution recirculated back to the fish. For horticultural farmers, it is estimated that the use of this innovative farming methodology requires as little as 95% less water than conventional farming.

7.3.4. Using the Sun to Generate Freshwater

A new trend in agriculture is the growth of greater productivity of using water for cultivation and increasing the food supply. There are no reliable long-term projections that show how much global population increase and climate change would affect the expansion will affect conventional and urban irrigation.

The idea that nearly 70% of the earth's surface is covered in water can be justified by the use of water shortages because its freshwater supplies are so vulnerable, makes this a strong case for desalination reasonable. Prior to that, however, the relentless need for electricity and the unreliability of profit margins have provided a stumbling block for companies that work with vast volumes of resources.

These systems are being used by Farms to harness solar power for the generation of energy for desalination of seawater to be used in hydroponic greenhouses.

The demand for farming is made without water, which is fed by fish farming and methods that draw no freshwater from the fossil or thermal water and use their fossil fuels, will create 300 jobs in Port Augusta, South Australia, for 10 years.

7.3.5. Agriculture by Aircraft

The vast amount of controversy about drones has to be seen in the mainstream media. It has been wrongly believed that drones were mostly used in the military and spying, but may be applied to more constructive uses, such as assisting farmers in routine flight and pizza delivery viral video projects too.

On top of this, the dark reputation linked with drones, there is also a competition each year that highlights how dynamic and, on the move, technology is in the Emirates of the UAE, called the UAE Drones for Good Award. Competition finalists have made a case for unmanned aerial vehicles (unmanned vehicle) aircraft as well as a method for environmental protection and a possibility for delivery of medications and machinery, for that year's prize.

It is like a traditional jet but has the capabilities of a multicopter, as well as the ability to allow vertical take-off and rapid flying, close to that of take-off, at high speed fixed-wing versions. It was developed by Munich-based Quantum-Systems. The usage of a quantum VRT design helps farmers to take advantage of fertilization plans that have environmental conditions and crop measurements, and outcomes overlayed, thereby reducing their dependency on fertilizers and increasing yields.

Dubai is planning to improve its food self-sufficiency by using agricultural drones by 30% to ensure food stability in the coming 30 years. As of this period, the emirate is importing 98% more than Singapore, its rivals, which means it is currently ahead of them (Figure 7.6).

Figure 7.6: Use of quadcopter in farming.

Sources: Image by Flickr.

7.3.6. Laser Scarecrows

Farming can be dated back to the start has not only to prehistory, but has persisted to modern times in many cultures, countries, and amongst bird advocates and farmers, including the one in America, where it started. Loss of revenue can be exacerbated by a percentage depending on pest bird populations, including starlings, blackbirds, which can eat up to 75% of the yield in the last 2 days of the harvest and cause major destruction within a short period.

The Pesticides have been used in all manner of getting rid of pests, including poisonous sprays and refined oils. From periodicals that remain conventional scarecrows to propane cannons, no ideas have been liked by natural selection.

The new deterrence developed in the past year at the University of Rhode Island keeps birds at bay. As opposed to visible sunlight, the laser light ray trainer projects green laser light that is invisible to humans. Since birds are so sensitive to green, the color green means something more.

Using a state-the-of-the-the-the-the-art laser system, the dart cannon will reach ranges of 600 miles. It causes such disturbance in flight that birds cannot damage the crops. Another advantage of this is it is less damaging to the ecosystem and less onerous on staff, which is why they are better.

Laser technology has been developed by various groups in different countries, with some employing the ability to aim and auto-track; there are some who use the power of the sun and those that use microwaves as a guide.

A far larger number of farmers and pest control specialists have taken it up as well, believing that with the invention, as much as 90% of their crops could be covered. It is a nonconfrontational, no burdening-enhancing action that does not bother the neighbors. It is phenomenally successful, with no risk of birds adjusting to its scare tactics to date.

7.3.7. Bee Vectoring

The disease prevention program has made a significant breakthrough with the recent finding of bee vectors. Bee vectoring has been shown to be useful in combined disease prevention and pest control by a corporation in Canada. This new system uses naturally occurring fungi called BVT-CR7, or Vectorite, which helps combat diseases that include wheat, barley rootworm and fungus.

The Vectorite solution is stored inside a hive of commercially raised bees, and as the bees leave the hive, they absorb the solution. The cure is then spread across the nearby crops and fields by the bees. In addition to improved disease tolerance, the technology improves plant growth and automatically extends grain shelf life.

Without the use of chemical pesticides, all of these advantages are available. Since it combines nature with agricultural industry processes, the invention is groundbreaking.

7.3.8. Harvest Quality Vision

The Harvest Quality Vision is a modern technology that eliminates the need for manual inspections during the harvesting process of fruits and vegetables.

HQV uses computerized scanners released this year to capture and assess crop quality and quantity. The enormous amount of berries and apples collected at a time makes it possible for the majority of fruits not to be classified prior to being stored.

When you use a camera to grade produce, there is no way to harm it by human or mechanical handling, and this process produces a greater quality than by removing defects. A warning will go off if it finds a disease, or an excess of one or shortage of specific inputs to the process, causing growers to change their behavior earlier rather than later in the process.

7.3.9. Product Testing In-Field or On-Site

Many crops such as Cannabis and industrial hemp must be tested rigorously and are subject to reporting requirements, whereas others, such as poultry, are less tightly regulated and can be tested in the field. Other crops benefit from being closely monitored for moisture and pollutant levels over their entire growth cycle.

In addition, these extra costs have to be absorbed for testing and lab procedures are associated with the need for further sampling. To lower the cost, hassle, extend in-field product research technologies call for innovative product production strategies that keep producers constantly making changes before a final review has been performed before the finished goods are produced.

Other light and portable equipment may provide a precise measurement of the total cannabinoid concentration that is not possible in a laboratory environment. Clipping and testing can now happen outside or in the field, making it possible for farmers to complete both quicker harvesting and for time and streamlining the production process.

The equipment may provide precise temperature information from which to provide a wider range of benefits, such as improved grain-temperature precision and accelerated analysis. As well as drying processes being regulated, the pricing paradigm is also extended, thereby eliminating charge-back and storage costs.

7.3.10. Crop and Soil Monitoring and Management

Extensive data collection and control are essential components of precision farming. It is time-consuming and susceptible to human error. Few of the companies that have achieved popularity for using sensors, drones, and satellite equipment to help farming. This type of technology can generate accurate images of the farm status, which can be transmitted through a cloud-based model to let farmers know how their operation is doing at all times.

Some applications are capable of preserving the soil such as agricultural processing. They can allow farmers to be aware of and explore and appreciate the minerals, leaves, and insects as well as their crops and cultivate more effectively and affordably. This type of sensor technology provides accurate and fast insights into agricultural activities that provide good and easy data flow.

An expanded list of nutrient distribution tools such as this and enhancement of environmental quality will help you formulate a more effective production strategy and improve yield, while still fine-tuning the method of application.

This is a successful procedure for accurate sampling, measuring, and plotting. Additionally, it is aided by GPS (global positioning system) and GIS. These soil-based and chemical steps now allow farmers to regulate the sum of different nutrients, the correct and salinity, and salinity as well as macro and chemical structure to enable them to meet optimum conditions while planting.

7.3.11. RFID

Radio-frequency identification (RFID) is a method that is used to tag a target with information that is reflected off of a conducting media. Though barcodes have typically been useful in a business environment, complex is more able to be read and scanned further away from the ground and incorporate a large amount of data because they are non-optical.

As with barcodes, RFID can be reprogramed, it is possible to store and display the full content of an expandable card, whilst it retains its ability to carry and expand up to 2 KB of data, including information such as position, date, and time.

By the amount of data precision, it decreases the probability of data mistakes while also improving the ability to detect and fix issues as they occur. There is less chance of human error for the enterprise systems that are supported by technology. People will be able to trace where their products are sourced, which will result in improved customer satisfaction and confidence.

7.3.12. Real-Time Kinematic (RTK)

Recently, a cutting-edge approach called real-time kinematic (RTK) has been used to improve the precision of current GPS signals. It has a comparable precision that brings to a GPS-level to only beyond the bulk of the normal error margin.

Radio-transmitted navigational satellites (RTK) are more precisely located and mapped in the field to estimate crop yields and expenses as compared to conventional satellites, as well as adjust expenses, making them more useful for accounting. A great deal of attention is paid to farm planning and plant density, with an approach for determining where the plants should be ideally situated and how they should be handled, is underway.

7.3.13. Vertical Farming

There are more and more people living in urban areas, industrialized countries that have less space to produce food, and this trend poses a danger for agriculture. About half of available arable land has already been wasted in the last four decades, and vertical farming could be the best way to stop it from getting any worse.

Because the process of cultivating plants in layers was extended, they can be generated in narrower spaces, for example, on vertical surfaces built into existing structures or in structures that already on the ground.

This will provide a much-needed solution to dealing with the escalating food needs as well as rising pressure on our resources, particularly in the city. Integrating greenhouses in this manner allows farmers to increase their income by using both the ecosystem surrounding them and the environment in the tower as one of their key factors for farming development.

Since the atmosphere changes very slowly over the year, this means the ability to grow year-round crops, with few negative impacts on yield (Figure 7.7).

Figure 7.7: Vertical farming of vegetable.

Sources: Image by Wikimedia.

7.3.14. Minichromosome Technology

In a recent example of genetic modification, engineers' mastery of mini-chromosomes has enabled the introduction of further-enhanced traits to an existing plant. Although it has extraordinarily little genetic material, the Minichromosome can be manipulated using genetic modification to give a biofortified or to make a dietary supplement. Analysis has shown that genetic modifications also boost crop tolerance, for example, the capacity to survive drought.

Since they contain only small quantities of genetic material, these Minichromosomes can easily be used to introduce foreign genes, without causing conflict with the normal development and growth of the host A complete understanding of the technique has not been achieved, but it has a lot of promise for various uses in agriculture. Genetically engineered crops are typically less negative in consumers' minds than in contrast to those grown from the plant's DNA, which has no disparaging baggage tied to it.

7.4. CONCLUSION

The transition to industrial agriculture is forcing farmers to grow increasingly bigger and produce more commodities to justify raising farm business expenses, reducing productivity, and putting at risk the local ecosystem's ability to support the farm's long-base businesses.

Even through farmers at all times have had trouble to thinking about it, there has never been a moment when they did not have anything to be worried about. Still, just as things have been successful and unsuccessful in the recent past, so will they be in the future. Additionally, even in these days when the land and livestock and farm crops were robust, there was still the threat of floods, plagues, droughts, or wars to fear.

In addition, farmers are grappling with everything of course with those ever-present problems. On the other hand, the current historical circumstances are different in certain ways. Farmers are additionally being challenged by the elements of the climate, as well as the economy has become more and more competitive over the years. Also, we are currently in the middle of an ongoing public health epidemic that is affecting nearly every part of our culture.

REFERENCES

1. (2019). *Innovations to Feed the World* (p. 2). [eBook] FAO. Available at: http://www.fao.org/3/ca3606en/ca3606en.pdf (accessed on 5 May 2021).
2. Croptracker.com. (n.d.). *Croptracker-10 Emerging Innovations in Agtech*. [Online] Available at: https://www.croptracker.com/blog/10-emerging-innovations-in-agtech.html (accessed on 5 May 2021).
3. Gupta, V., (2019). *Challenges of Seed Production in India*. [Online] Businesseconomics.in. Available at: http://businesseconomics.in/challenges-seed-production-india#:~:text=Among%20various%20challenges%20faced%20by,heavily%20in%20pesticides%20and%20fertilisers (accessed on 5 May 2021).
4. Jagranjosh.com. (2014). *Problems Related to Irrigation*. [Online] Available at: https://www.jagranjosh.com/general-knowledge/problems-related-to-irrigation-1409136420-1 (accessed on 5 May 2021).
5. Jha, R., (2005). *Land Fragmentation and its Implications for Productivity: Evidence from Southern India*. [Online] Research Gate. Available at: https://www.researchgate.net/publication/5223785_Land_Fragmentation_and_its_Implications_for_Productivity_Evidence_from_Southern_India (accessed on 5 May 2021).
6. McClelland, J., (2015). *Top Five Technology Innovations in Agriculture*. [Online] Raconteur. Available at: https://www.raconteur.net/sustainability/top-5-tech-innovations-in-agriculture/ (accessed on 5 May 2021).
7. Mondal, P., (n.d.). *Ten Major Agricultural Problems of India and their Possible Solutions*. [Online] Your Article Library. Available at: https://www.yourarticlelibrary.com/agriculture/10-major-agricultural-problems-of-india-and-their-possible-solutions/20988 (accessed on 5 May 2021).
8. Politico, (2016). *Innovative Agriculture: Challenges and Solutions for Sustainable and Competitive Food Production*. [Online] Available at: https://www.politico.eu/event/innovative-agriculture-challenges-and-solutions-for-sustainable-and-competitive-food-production/ (accessed on 5 May 2021).

9. Veeck, G., Veeck, A., & Yu, H., (2020). Challenges of agriculture and food systems issues in China and the United States. *Geography and Sustainability, 1*(2), 109–117.

Chapter 8

Impact of Climate Change on Agriculture

CONTENTS

Climate change is one of the most important factors that determine the overall success of the agriculture sector. In the chapter, impact on climate change on agriculture, the role of diagnosis and forecast of global climate change is discussed in which current conditions of global warming are evaluated. In addition, it also demonstrated how climate change will alter our food.

It also tries to answer question of how will global food production be affected in the future by climate change? It also stated how climate change will challenge farmers. In addition, it also discusses how to help farm communities in managing severe impacts.

Some indirect impacts of climate change on agricultural productivity are also discussed. It also shed some light on non-climate impacts related to greenhouse gas emissions. In the end, the chapter tries to explain how one can predict impacts of climate change impacts some of the tools for adaptation.

8.1. INTRODUCTION

Climate change can be defined as the changes that take place beyond an average atmospheric condition that is mainly caused by natural factors as well as artificial factors. The natural factors include volcanic activities, the orbit of the earth's revolution, crustal movements, etc., while on the other hand, the artificial factors include the increase in the concentration of aerosol and greenhouse gases (GHGs).

Climate change that takes place because of the effect of global warming refers to the average increase in the global temperature. Global warming has gradually become a megatrend that will eventually lead to significant global changes in the climate in the near future.

Related to such impacts, the UN Intergovernmental Panel on Climate Change (IPCC) presented significant scientific evidence in its fourth report on climate change, which has now become recognized all around the world. Further, now people have become more aware of the fact that global warming cannot be avoided because of the sustained increase in the emissions of GHGs and the changes in the climate system because of that.

Officially in the year 1972, the Club of Rome Report raised global warming as a global issue and, in year 1985, World Meteorological

Organization (WMO) and United Nations Environment Program (UNEP) officially declared carbon dioxide as the main cause of global warming.

To effectively deal with the issue of global warming, Intergovernmental Panel on Climate Change (IPCC) was organized in the year 1988 and has carried out in-depth studies as well as systematic research on climate change.

It is somehow indisputable that global warming has severe impacts on the earth. There are higher chances that GHGs emissions will increase more and more because of anthropogenic activities. Global warming not only causes a severe change in the temperature and precipitation, but it also increases the frequency of droughts, floods, heatwaves, the intensity of typhoons and hurricanes following the change in temperature and precipitation patterns.

In spite of all these changes, the impact of climate change can also be seen in various other forms all around the world. These could be a rise in sea level, northward movement of plant habitats, decrease in glaciers, rice in ocean temperature, changes in animal habitat, the early arrival of spring, and shorten winter.

The increase in global warming effects not only affects the ecological system but also human life. It has now become a major issue all around the world. The approaches by which the issue of global warming can be resolved or managed to some extent are divided mainly into mitigation measures, emphasizing on reduction and absorption of GHGs, the causative factors, and the adaptation measures to reduce the damages caused by climate change.

To agriculture, agricultural crop production is very much vulnerable to climate variability and, climate changes associated with an increase in carbon dioxide, increase in temperature and changing pattern of rainfall may ultimately lead to an extensive reduction in the production of crops.

In addition, extreme weather events like heatwaves, droughts, heavy rainfall, flood, etc., have increased in the past few years. Improving the production of crops to meet the increasing demands owing to the increasing population, against the background of the pressures of climate change, is a challenging task.

As a result, more attention is required for adaptation and mitigation research, change in policies, capacity building, regional or national cooperation, and the potential support of national or global adaptation funds along with certain other resources to minimize the severe impacts of climate change on crops.

Adaptation to climate variability and some of the extreme events serve as the base for minimizing vulnerability to a long-term climate change. Some of the simple adaptation practices like the use of climate-ready crops or thermal stress-tolerant varieties, improving water conservation and management practices, adjusting planting dates, diversifying crops, improving pest management, and using efficient irrigation and fertilizer management could efficiently aid in reducing the severe impacts of climate change.

8.2. DIAGNOSIS AND FORECAST OF GLOBAL CLIMATE CHANGE

Figure 8.1: Impact of climate change on food productivity.

Source: Image by Food Business News.

8.2.1. Current Conditions of Global Warming

Climate is a long-term variation in the atmospheric conditions of a particular region. On the other hand, climate change refers to the steady change in the climate system both by natural as well as artificial causes. The key cause behind the change in climate is the change that takes place on each component of the climate system such as atmosphere, hydrosphere, biosphere, cryosphere, and lithosphere or by complicated interactions among those components.

In addition, the causes of climate change are mainly divided into:

- Natural causes; and
- Artificial causes.

Natural causes further include the change in solar activity, seawater temperature, volcanic eruption, westerly waves, atmospheric waves, and ice cap distribution. Whereas, artificial causes include emissions of carbon dioxide from agricultural activities, industries, acid rain, deforestation, and the destruction of the ozone layer by Freon gas, with global warming by the increase of greenhouse gases as the representative (Presidential Advisory Council on Education, Science, and Technology: PACEST, 2007).

Global warming is referred to the average increase of the temperature of the earth because of the greenhouse effect caused by carbon dioxide (CO_2), nitrous oxide (N_2O), methane (CH_4), hydrofluorocarbon (HFCs), perfluorocarbon (PFCs) and sulfur hexafluoride (SF_6).

Global warming which is known as the cause behind the increased temperature of the earth started initially from the time of the Industrial Revolution, which was accompanied by a quick increase in fossil fuel consumption. That issue attracted global interest as the scientific knowledge of climate has accumulated since the period of 1970s. In addition, it has been largely accepted by scientists and researchers that anthropogenic greenhouse gas emissions are the cause of global warming.

The global concentration of greenhouse gas based on carbon dioxide is estimated to have increased from 280 ppm before the Industrial Revolution (1750) to 379 ppm in the year 2005. According to an analysis of the average temperatures of the earth done by the Climate Research Unit, in the year 2009, the increase of the earth's average temperature so far appears to be much higher than the increase before the Industrial Revolution.

More precisely, global warming has suggestively accelerated since the year 1980, and the average temperature of 1998 was observed to be 0.58°C higher than the average temperatures of 1960~1990.

8.2.2. Current Conditions and Forecasts of Global Climate Change

As global climate change is affected by different factors like regional characteristics, socioeconomic variables, and meteorological variables, it is projected by making various feasible scenarios. The IPCC Assessment

Report provides greenhouse gas emission scenarios by projecting the change in greenhouse gas concentration as per the demographics and socioeconomic development.

The Special Report on Emission Scenario (SRES) presents four main scenarios (A1, A2, B1, and B2) and three other scenarios (A1F, A1T, A1B) modified as per their technological emphasis in the A1 scenario (Figure 8.2).

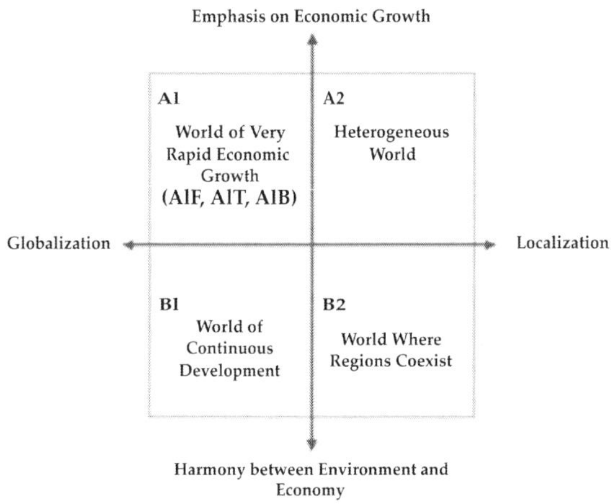

Figure 8.2: Conceptual diagram of scenarios for estimating climate change.

Source: Image by Kim, Chang-Gil et al.(2009), p. 21.

The A1 scenario assumes a very rapid economic growth, in which the rapid growth of the global economy and population peaks in 2050 and declines thereafter, in which then-new efficient technologies are introduced.

According to the alternative development of energy technology, it is further divided into three groups. The three scenarios are the fossil intensive scenario (A1F1), non-fossil energy scenario (A1T), and balanced-energy source scenario (A1B). A2 is the scenario for a heterogeneous world with a high rate of population growth, a low economic growth rate, and the most varied but gradually developing technologies.

The B1 scenario assumes the same population growth rate as that of the A1 scenario. However, BI scenario assumes a lower economic growth rate. In this case, the economic structure changes toward a service and information economy and sustainable development are pursued with stress on clean as well as resource-efficient technologies.

B2 is a scenario for a world where regions coexist with each other in harmony. This scenario assumes the intermediate level of population and economic growth between A1 and B1, and focuses on regional solutions for social, economic, and environmental sustainability.

8.3. HOW CLIMATE CHANGE WILL ALTER OUR FOOD?

It has been predicted that by the end of the year 2050, the world population would grow to around 10 billion. With around 3.4 billion more mouths to feed, and the increasing desire of the middle class for dairy and meat in the developing nations, the global demand for food could increase by between 59%and 98% (Figure 8.3).

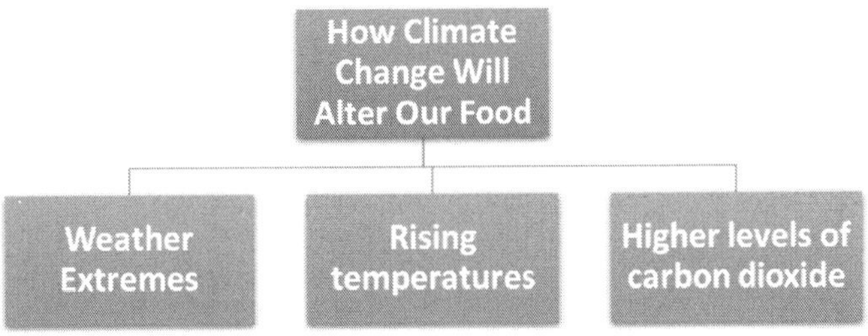

Figure 8.3: How climate change will alter our food?

This means that agriculture, all around the world, needs to improve its production as well as increase yields. However, according to scientists, the impacts of climate change such as extreme weather, higher temperatures, increasing levels of carbon dioxide, sea-level rise, flood, drought—threaten to decline the quantity and threaten the quality of our food supplies.

A recent study of global vegetable and legume production determined that if GHGs emissions continue as they are on the current trajectory, then the yields would fall by 35% by the year 2100 because of increased salinity, ozone, and water scarcity.

Another study found that US production or corn which is also known as maize, most of which is used for feeding the livestock and produce biofuel, could be cut in half by a 4°C increase in global temperatures—which could happen by 2100 if there will be no reductions in our greenhouse gas emissions.

If the warming is limited under 2°C, the goal of the Paris climate accord, U.S. corn production could still decrease by about 18%. In addition to that, researchers also found that the risk of the world's top four corn exporters that are U.S., Brazil, Argentina, and Ukraine, suffering concurrent crop failures of 10% or more is about 7% with a 2°C increase in temperature. If temperatures rise to 4°C, the odds shoot up to a staggering 86%.

"We're most concerned about the sharply reduced yields," said Peterde Menocal, Dean of Science at Columbia University and director of the Center for Climate and Life. "We already have trouble feeding the world, and this additional impact on crop yields will impact the world's poorest and amplify the rich/poor divide that already exists."

However, the change in climate will not only affect crops—it will also impact meat production, fisheries along with several other fundamental aspects of food supply.

8.3.1. Weather Extremes

Around 80% of the total crops of the world are rainfed. Thus, most of the farmers depend largely on the predictable weather that agriculture has adapted to produce their crops. Although, climate change is gradually altering the patterns of rainfall all around the globe.

When temperature increases, the warmer air holds much more moisture and thus, it can make the precipitation much intense. Events of extreme precipitation that are becoming very common, can damage the crops directly that would eventually result in a reduction of yields.

Flooding that results from the growing intensity of the tropical storms and rise in sea level is also likely to increase with the change in climate, and ultimately can drown crops. As floodwaters can transport manure, sewage, or pollutants from roads, lawns, and farms, more and more toxins and pathogens could find their way into the food.

Hotter weather will eventually lead to faster evaporation. This will result in water shortages and more droughts, and thus there will be very little water available for irrigation purpose. Around 10% of the total crops grown all around the world in their major regions of food production are irrigated with groundwater which is non-renewable.

Alternatively, it can be said that aquifers are beings drained at a much faster rate as compared to the rate at which they are refilling. This could be regarded as an issue which will get worsen as the world continues to experience high temperature.

This is taking place in major regions where food is being produced, such as the U.S. Great Plains and California's Central Valley, and in India, Pakistan, northeastern China, and parts of Iraq and Iran.

Depletion of groundwater is a slow-building pressure on the food system. Countries do not have any effective policies in place that can deal with the fact that major resources are being depleted continuously by humans, especially in food-producing regions, and this is disconcerting.

According to the climate projection, drought will become very common in much of the US parts, mainly the southwest. In some other parts of the world, water shortages and drought are expected to impact rice production, which is a staple food for more than half of the people on the Erath.

At the time of severe drought years, rainfed rice yields have been decreased from 17% to 40%. In South and Southeast Asia, around 23 million hectares of the rainfed rice production areas are already experiencing water scarcity, and periodic drought affects around 80% of the rainfed rice-growing areas of Africa.

Extreme weather conditions such as drought and heavy storms can also damage the transport of food. This could also increase the chances of spoilage as well as contamination unless the food is stored properly. This would eventually result in many foodborne illnesses.

8.3.2. Rising Temperatures

Global warming may however benefit some of the crops like potatoes mainly in Northern Europe and rice in West Africa. This could also make some of the farmers raise new crops that only grow in warmer areas (Figure 8.4).

Figure 8.4: Rising temperature causing drought.

Source: Image by weforum.org.

The eventual effect of increasing heat depends largely on the crop's optimal range of temperatures for growth as well as reproduction. If the temperatures exceed their optimal range, then the yield will ultimately drop as the heat stress would damage the plant's pollination, root development, flowering, and other growth stages.

As per a National Academy of Sciences report, for every degree Celsius that the global thermostat increases, there will be a 5 to 15% reduction in overall crop production. Heat waves that are expected to become much more frequent, make livestock less fertile and thus, much more vulnerable to diseases. It is a fact that dairy cows are really sensitive to heat, as a result of which milk production could decline.

Parasites and diseases that target the livestock, usually grow in warm and moist conditions. This could ultimately result in livestock farmers treating diseases of animals and parasites by using veterinary medicines and different chemicals, which might also get into the food chain.

In addition, climate change will also enable pests, weeds, and fungi to increase their range as well as numbers. Additionally, milder winters and earlier springs will allow more of such pests and weeds to survive for a long period of time.

Pests and plant diseases that are novel to an area could destroy crops that have not had time to evolve defenses against them. For instance, novel virulent mutant strains of wheat rust, a kind of fungal infection that had not been observed for over 5 years, have spread from Africa to Asia, the Middle East and Europe, devastating crops.

8.3.3. Higher Levels of Carbon Dioxide

As plants use carbon dioxide to prepare their food, more carbon dioxide in the atmosphere can improve the yields of the crop in certain areas if other conditions such as soil moisture, nutrient amounts, and availability of water are perfect. However, the beneficial effects of increasing levels of carbon dioxide on plant growth can be offset by severe weather, heat stress or drought.

As higher levels of carbon dioxide can stimulate the growth of plants as well as increase the number of carbohydrates produced by the plants, this comes at the cost of vitamin, minerals, protein content. Researchers found that the protein content will likely decrease gradually if the level of carbon dioxide reaches 540 to 960 parts per million, which is projected to reach by 2100. At present, the level of carbon dioxide is at 409 ppm).

Several studies show that wheat, barley, rice, and potatoes have almost 6 to 15% lower protein concentration when they are grown at those levels of carbon dioxide. However, the protein content of corn and sorghum did not significantly decline.

In addition to that, the concentration of certain important elements, for example, zinc, iron, magnesium, calcium, sulfur, copper, nitrogen, and phosphorus, are expected to decline with more and more presence of carbon dioxide in the atmosphere. When the level of carbon dioxide rises, the openings in plant shoots and leaves shrink, as a result of which, plants lose water. Various research work recommends that as plants lose water more slowly, their circulation eventually slows down, and thus they draw in a very little amount of nitrogen and minerals from the soil. Levels of vitamin B in crops may also decline, as nitrogen's presence in plants is important for producing vitamins.

8.4. HOW WILL GLOBAL FOOD PRODUCTION BE AFFECTED IN THE FUTURE BY CLIMATE CHANGE?

Agriculture is a key source of GHGs, which in turn contributes to the greenhouse effect as well as climate change. Though, the changing climate is having very harsh impacts on agricultural production that will challenge food security in the near future.

Fundamentally, global food security depends on enough production of food as well as easy access to food. According to FAO (1996), food security is defined as a state when "all people, at all times, have physical and economic access to sufficient, safe, and nutritious food to meet their dietary needs and food preferences for an active and healthy life."

At present, the main barrier in the path of food security is food access. Enough amount of food is being produced all around the world to feed the population, but somehow more than 10% are undernourished.

There are estimates that climate change will significantly contribute to food insecurity in the coming years, by increasing the prices of food and reducing its production. As climate change mitigation efforts increase energy prices, food may become very expensive in the future.

Water needed for the production of food may become very limited because of the increase in water usage for crops, and also, a big reason could be drought. Also, the competition for land may increase as some of the land areas may become climatically not suitable for farming or production.

Additionally, extreme weather conditions related the climate change may eventually cause a sudden decline in agricultural productivity, and thus leading to rapid price increases.

Heatwaves, for example, in the summer of the year 2010 led to yield losses in major areas of production such as Ukraine, Russia, and Kazakhstan and contributed to a huge increase in the price of staple foods. These increased prices forced increasing numbers of local people into poverty, providing a sobering demonstration of how the impact of climate change can eventually cause food insecurity.

The agreement of the Intergovernmental Panel for Climate Change (IPCC) is that considerable climate change has already taken place since the 1950s, and there are chances that the global mean surface air temperature will increase by 0.4 to 2.6°C in the future (depending on future greenhouse gas emissions).

Agriculture being a wider food production system is a major source of GHGs emissions. Future intensification of agriculture for compensating the reduced food production due to climate change, along with an increasing food demand could further increase such emissions. It is estimated that the demand for livestock products will grow by +70% until 2050.

A gradual rise in the level of carbon dioxide and temperature may eventually result in favorable conditions, in turn increasing the yields of certain crops. On the other hand, in some regions, such potential yield increases are likely to be limited by severe events such as flood, drought, or extreme heat, during the period of crop flowering.

Production of crops is estimated to decline in various areas because of the severe climate changes. Figure 8.5 summarizes the projections of average crop yield across all emission scenarios, regions, and with- or without-adaptation by farmers, showing an increasing trend towards widespread yield decreases.

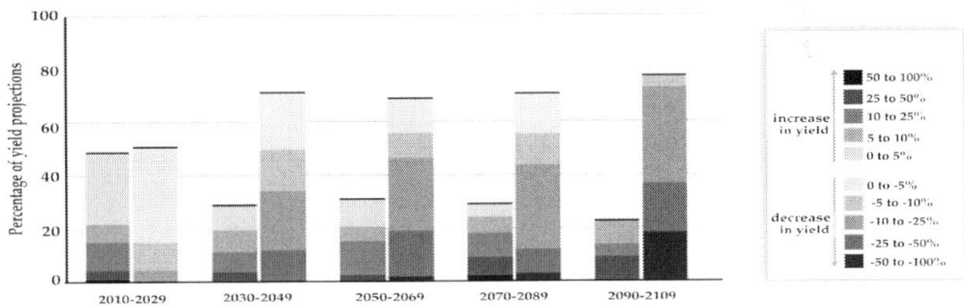

Figure 8.5: Summary of projected in crop yields, due to climate change over the 21st century.

There are high chances that heat waves that are known as periods of extremely high temperature will become very frequent in the near future and would be a key challenge for agricultural practices. Heatwaves can also cause heat stress in both plants as well as animals. It has a negative impact on the production of food.

It is evident that the extreme period of high temperatures is very harmful for the production of crops if they occur at the time of crop flowering. If this critical stage of flowering is disrupted, there would be no seeds at all. Also, the animals get highly affected by heat stress. It can result in lower fertility and lower productivity in animals. This can also cause negative effects on the immunity and make them prone to diseases.

The impact of heatwaves is expected to be non-uniform, with strangely negative effects in less developed nations. Along with other aspects of climate change like increased drought intolerance, they may even make worse the current issues revolving around food security.

The estimated changes in climate are not only restricted to an increase in temperature and heat waves, a major change in rainfall patterns is also projected to take place. As some of the regions are likely to suffer more from droughts in the future, there are some other regions that are expected to deal with certain opposing issues of increased flooding due to torrential rains. In coastal areas, rising sea levels may eventually result in complete loss of agricultural land.

Warmer climates may ultimately lead to more severe issues from pests and diseases, along with shifts in the geographical distribution of certain pests. For instance, insects that serve as a vector for transmission of diseases are likely to further migrate poleward in the future, where livestock have not been exposed to such diseases so far.

The responses of yield to different kinds of stresses have been well described by experimentation in different crops. Quantification of such responses and recognizing the phase when agriculture is most vulnerable to stress is important for identifying the most efficient strategies for adaptation.

Crop-level adaptation to change in climate is expected to be the main cause behind the decline in future yield losses, and it may involve changing crop cultivars, cultivation techniques, sowing time, and/or irrigation practices. Ongoing research is addressing the issues or challenges of maintaining and/or increasing crop production under global change.

Some risks to the production of crops from climate change and extreme weather conditions have been recognized and effective strategies are recommended to help maintain the food production. The strategies include: restoring farm type, crop, or cultivar scale diversity into food systems, to improve their resilience and making crop improvements that enhance stress tolerance.

In addition to that, other strategies may include developing pre-defined, international responses to food shortages to prevent food price shocks that might reduce people's access to food.

8.5. HOW CLIMATE CHANGE WILL CHALLENGE FARMERS?

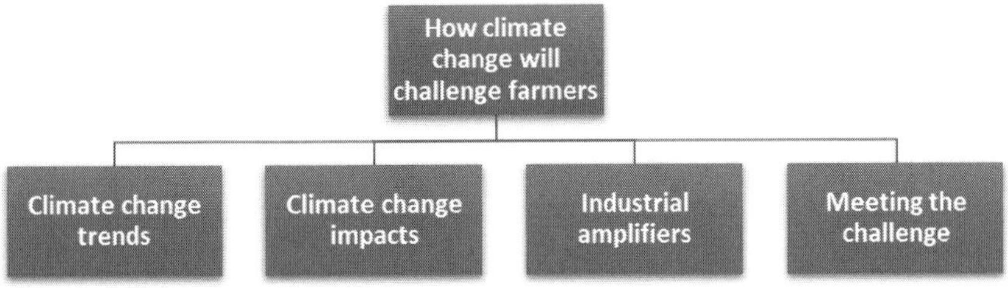

Figure 8.6: How climate change will challenge farmers?

8.5.1. Climate Change Trends

Changing precipitation patterns: Pattern of rainfall have already started shifting across the country. Such changes are expected to be increased over the coming years. There are high chances of more intense periods of heavy rainfall and longer dry periods, even in the same region.

Changing temperature patterns: Increase in average temperatures and extreme heat throughout the year, lesser cool days during winters, and more frequent cold season thaws will affect farmers of each and every region.

8.5.2. Climate Change Impacts

Floods: In the past few years, an increase in flooding in many agricultural regions are observed. The rise in sea level too is intensifying the intensity as well as the frequency of flooding on farms in coastal areas. Such costly floods damage the crops and livestock, too, along with accelerating soil erosion, polluting water, and damaging schools, roads, and various other infrastructure.

Droughts: Both too much as well as too little water is damaging for agriculture. Extreme droughts have taken a heavy toll on crops, farmers, and livestock in different regions over the past decade. According to science, an increase in temperatures would even make droughts worsen and deplete water supplies and, in some cases, spurring destructive wildfires.

Changes in crop and livestock viability: Farmers choose such crop varieties that are very well suited to local conditions. As those conditions change gradually over the coming years, a lot many farmers will be forced to rethink some of their preferences-that can mean making new capital investments, learning new practices, and finding new markets.

New pests, pathogens, and weed problems: As farmers will require to find new crops, practices, and livestock, they will have to deal with new issues and challenges. A weed or an insect that could not thrive north of Texas in decades past may find Iowa a perfect fit going forward, and farmers will have to adapt.

8.5.3. Industrial Amplifiers

Degraded soils: Typical systems of monoculture cropping leave soil bare for much of the year, depend on synthetic fertilizer and plow fields on a regular basis. Such practices leave soils low in organic matter and prevent the formation of deep and complex root systems. Amongst the results, reduced water-holding capacity (which worsens drought impacts), and increased susceptibility to erosion and water pollution (which worsens flood impacts) (Figure 8.7).

Figure 8.7: Industry contribution in uncertain climate conditions.

Source: Image by Uscusa.org.

Simplified landscapes: Industrial agriculture treats the farm as a crop factory instead of a managed ecosystem, with minimal biodiversity over a large area of land. The lack of diversity in the operation of farming exposes farmers to high risk and increases the climate impacts like changes in crop viability and encroaching pests.

Intensive inputs: A heavy reliance of industrial farm on fertilizers as well as pesticides may become very costly for struggling farmers as climate impacts increase soil erosion and various pest problems. Heavy use of chemicals will also increase the pollution faced by downstream communities as flooding increases. Farmers may also increase irrigation in response to increasing temperature extremes and drought, and exhausting precious water supplies.

8.5.4. Meeting the Challenge

Business, in general, would not protect the future of food supply or the well-being of the farmers as well as communities that produce it. Concrete steps are required for the preparation of dealing with climate impacts on agriculture and to decrease both their severity and our vulnerability to them.

Climate change risks are not equally distributed, and neither are the pathways to the adaptation of climate. Public as well as institutional policies have long denied the access of low-income groups, communities of color and tribal communities to important resources and critical decision-making processes, leaving them with lesser options and much more risk in the name of climate impacts. Thus, it is important to guarantee that such communities have a voice in shaping effective adaptation strategies.

8.6. HELPING FARM COMMUNITIES MANAGE SEVERE IMPACTS

When the impacts of climate change strike, support systems require to be in its effective place to help the communities cope as well as recover. Shelters and certain other facilities to provide housing, first aid, food, and other immediate needs for people whose lives have been severely disrupted by fires, droughts, floods, or storms (Figure 8.8).

Figure 8.8: Helping farm communities manage severe impacts.

Investment in the local capacity as well as infrastructure to support people that are affected by the climatic impacts as they rebuild their lives and businesses too. This not only includes the infrastructure for communication, transportation, water, and sanitation but also training in novel practices as well as opportunities that build adaptive capacity.

8.6.1. Reducing Damage by Making Farms More Resilient

Farms and farm communities do not require to be sitting ducks for the climate impact. Scientists and forward-looking farmers are finding effective climate-resilient ways of producing food. Build healthier and spongier soils via practices like planting cover crops as well as deep-rooted perennials that can eventually increase the capacity of soil to soak up heavy rainfall and hold water for dry periods.

It is important to make farms stronger by redesigning them as diverse agroecosystems, incorporating trees, native perennials, reintegrating crops, and livestock, and reducing dependence on pesticides and fertilizers, developing new varieties, livestock breeds, and farm practices precisely designed to help farmers adapting evolving climate realities.

8.6.2. Addressing the Root of the Problem

Eventually, whatever is done for helping farmers to adapt climatic changes, still they face urgent needs and obligations for reducing the source of the issue. It means bringing net emissions of heat-trapping gases down to zero. Luckily, farm and farm systems can be the most important part of the solution. They can reduce the emissions at every stage of the production as well as distribution process of food, and also by building such agro-systems that can store more carbon.

8.6.3. Policy Recommendations

What policy levers can be pulled up to help get these solutions off the ground?

- **Invest in Public Research:** Research that is publicly funded provides farmers with the tools as well as information that they require to increase productivity along with efficiency. Due to climate change and its impacts, farmers now need science more than ever, yet public funding for research can help them deal with conditions of short supply.

Agroecology research, research that produces the kind of long-term, and root-deep solutions that can effectively help farms stay viable for generations.

The expansion of conservation programs in the federal farm bill will make it easier for the farmers to adopt effective and sustainable practices. These will ultimately make their farms climate-resilient. Their funding needs to be boosted, along with their impact.

- **Strengthen Safety Nets (and Make Them Drivers of Resilience):** Irrespective of what science and progressive policies can do, farms all around the world will be challenged, some more than the others. It is important to provide farm communities and families with the potential support that they require to survive the climate crisis and eventually become much more resilient.

This further consists of crop insurance programs, effective, and responsive disaster relief programs, and health care access for farmers and farmworkers. To attain net-zero emissions of hazardous gases, it is required to prioritize the policies to decrease climate emissions in a large amount.

8.7. INDIRECT IMPACTS OF CLIMATE CHANGE ON AGRICULTURAL PRODUCTIVITY

8.7.1. Pests and Diseases

An increase in atmospheric carbon dioxide and climate change may indirectly impact the crops affecting pests and diseases. Such interactions are really complicated and still complete implications in reference to crop yield are uncertain.

Indications recommend that pests, such as aphids (Newman, 2004) and weevil larvae (Staley and Johnson, 2008), respond positively to increased carbon dioxide. In addition, rise in temperatures also declined the overwintering mortality of aphids enabling earlier and potentially more widespread dispersion (Zhou et al., 1995).

Also, evidences suggest that in sub-Saharan Africa, the migration pattern of locusts may be highly influenced by the patterns of rainfall (Cheke and Tratalos, 2007) and as a result, potential exists for climate change that can shape the impacts of this disturbing pest. Pathogens as well as disease may also be affected by a change in climate (Figure 8.9).

Figure 8.9: Indirect impacts of climate change on agricultural productivity.

Source: Image by unsplash.com.

This may be possible through the impacts of drought or warming on the resistance of the crops to particular diseases and also, through the increased level of pathogenicity of organisms by induced mutation caused by the environmental stress (Gregory et al., 2009).

Over the next one or two decades, disease that affect oilseed rape could increase in harshness within its existing range and thus, will spread to more northern regions where at present, it is not observed (Evans et al., 2008). Changes in climate variability may also prove to be critical, affecting the predictability as well as amplitude of outbreaks.

8.7.2. Changes in Water Availability Owing to Remote Climate Changes

The climatic changes that happen at a distant region affect agricultural production to a large extent. For instance, the agriculture which takes place along the Nile, Egypt is completely dependent upon the rainfall in the upper reaches like Ethiopian Highlands.

The transformation in the patterns of snow affects the store and release of water and the change in the quantity of precipitation impacts the run-off volume when the snow starts to melt in the end of the winter. Whereas the fluctuation in temperature has a large impact in the timings of run-off during spring.

The water for irrigation is withdrawn from rivers that depend upon the climate of a distant place. Irrigated agricultural land produces 40–45% world's total food while taking 1/5 of cropped area.

When there is a capacity to store run-off during the times of abundance to avail it later in the growing season, then the additional river flow is considered favorable. Where storage capacities are not ample, there the winter run-off is instantly lost to the oceans.

Any change happening in precipitation far away as well as the magnitude of glacial meltwaters determines and impacts food production for approximately one-sixth of the world's population who reside in glacier-fed river basins. Some prominent rivers such as Indus and Ganges are fed by mountain glaciers.

The prevalence of observed glaciers around the world are compressing for various reasons and warming being the initial cause of retreat along with the presence of atmospheric moisture in the tropics.

In the future, the melting of glaciers will increase the river flow due to which the risk of flood will eventually increase as well as it will result in a decline in run-off, though the time scale is undetermined.

"The Chinese Glacier Inventory" listed up 46,377 glaciers in western China and 15000 glaciers in the Himalayas. About 12000 Km^3 of freshwater is stored in these glaciers. (Ding et al., 2006; Cruz et al., 2007).

The glaciers in the western Himalayas, when examined properly, denotes glacial thinning and glacial retreat, whereas radioactive bomb deposits from one high altitude glacier show no net rack up since 1950 (Kendall et al., 2008).

The retreating glaciers and water flows in the Himalayas cater immense rivers like Indus, Ganges, and Brahmaputra and contributes seasonal river flow in a salient proportion but its immensity is unspecified.

The climate changes that occur affects Indus and Ganges. Due to these changes, the flow of these rivers is halted in summers turning them into seasonal rivers. The fact that approximately 500 million people are dependent on these rivers, in the coming time, water scarcity will increase due to the demographic explosion.

8.7.3. Mean Sea-Level Rise

Rise in sea level is an unavoidable result of a warming climate, that owes to a combination of thermal expansion of the current mass of water present in the ocean, along with addition of extra water that owes to the melting of land ice.

In addition, this can be expected to cause inundations of the coastal land ultimately, mainly in the places where the capacity for introduction or modification of the sea defenses is non-existent orbit lower. In terms of crop productivity, vulnerability is obviously the greatest where the high rise in sea levels occurs in combination with low-lying coastal agriculture.

A lot of the major river deltas provide critical agricultural land owing to the richness of fluvial soils, and also, many small island states are low-lying. Increases in total sea level also causes threats to inundate agricultural lands as well as salinize the groundwater in coming years to decades to centuries. However, the major impacts may not be observed for so many centuries owing to the requirement of time needed to melt large ice sheets and also for warming to penetrate into the deep ocean.

The potential rise in the sea level is linked with the melting of the main ice sheets. For West Antarctic Ice Sheet (WAIS), it is 5 m, for East Antarctic Ice Sheet (EAIS), it is 60 m and for Greenland Ice Sheet (GIS), with both the GIS and WAIS considered vulnerable, it is 7 m.

So, because of the possible rate of discharge of these ice sheets and past maximal sea-level rise, a maximum eustatic sea-level rise of around 2 m by the year 2100 is considered physically plausible, but very unlikely (Pfeffer et al., 2008; Rohling et al., 2008; Lowe et al., 2009).

Short-lived surges in storm too, can cause huge devastation, even if the land is not lost permanently. Relatively, there has been very little work evaluating the impacts of either mean rise in sea-level or surges of storm on agriculture.

8.8. NON-CLIMATE IMPACTS RELATED TO GREENHOUSE GAS EMISSIONS: IMPACTS OF CHANGES IN ATMOSPHERIC COMPOSITION

8.8.1. CO_2 Fertilization

Along with impacting climate by radiative forcing, an increase in atmospheric carbon dioxide concentrations can also affect the plant physiological processes of photosynthesis as well as transpiration (Field et al., 1995). As a result of which any assessment of the impacts of CO_2-induced climate change on productivity of crops should account for the modification of the climate impact by the CO_2 physiological impact.

The physiological response of carbon dioxide differs between different species and especially in two different pathways of photosynthesis (C3 and C4) have evolved and affect the overall response. The key difference lies in whether ribulose-1,5-bisphosphate carboxylase-oxygenase (RuBisCO) within the plant cells is saturated by CO_2 or not (Figure 8.10).

Figure 8.10: Carbon dioxide fertilization.

In C3 plants, RuBisCO is not CO_2-saturated in current atmospheric conditions, so increasing concentrations of CO_2 increase the total uptake of carbon and as a result, growth. The RuBisCO enzyme is highly conserved in plants, and as such, it is thought that the response of all C3 crops, including wheat and soya beans will be comparable.

Certain theoretical estimates recommend that increase in CO_2 concentrations up to 550 ppm, could eventually increase the process of photosynthesis in C3 crops by around 40% (Long et al., 2004). The physiology of C4 crops, such as millet, maize, sugarcane, and sorghum are different.

In such plants, CO_2 is concentrated to 3 to 6 times of the atmospheric concentrations and therefore, RuBisCO is already very saturated (von Caemmerer and Furbank, 2003). Therefore, it is clear that the rising CO_2 concentrations confer no additional physiological benefits.

However, such crops may become more and more water-use efficient at increased CO_2 concentrations as stomata do not require to stay open for a longer period of time, for the plant to get the required CO_2. Hence, yields may marginally rise (Long et al., 2004).

Many studies recommend that yield increases owing to the fertilization effect of CO_2, and these consequences are consistent across a wide range of the experimental approaches such as controlled environment closed chambers, open, and closed field top chambers, greenhouse, and free-air carbon dioxide enrichment (FACE) experiments (Tubiello et al., 2007).

Experiments performed under ideal conditions show that a doubling of atmospheric CO_2 concentration increases photosynthesis by 30–50% in C3 plant species while, on the other hand, 10–25% in C4 species (Ainsworth and Long, 2005).

Increase in crop yield is lower than the photosynthetic response. An increase in atmospheric CO_2 to 550 ppm would increase C3 crop yields by 10–20% and C4 crop yields by 0–10% on an average (Gifford, 2004; Long et al., 2004; Ainsworth and Long, 2005).

Some of the authors argue that the response of crop to increased level of CO_2 may be lower than earlier, with effects for crop modeling as well as projections of food supply (Long et al., 2004, 2009). Plant physiologists recognize that the effects of increased CO_2 as calculated in experimental settings and eventually executed in models may overestimate the actual field and farm level responses.

This is mainly because of various limiting factors such as weeds and pests, nutrients, competition for resources, soil water, and quality of air, which are neither well understood at large scales, nor well-executed in leading models.

In spite of the potential positive effects on the quantities of yield, increased concentration of CO_2 may be detrimental to the quality of yield of certain crops. For instance, increased CO_2 is harmful to the quality of wheat flour through reductions in protein content (Sinclair et al., 2000).

Global-scale comparisons of the effects of CO_2 fertilization with those of changes in climatic conditions show that the potential of CO_2 fertilization effects is the most critical factor in defining whether the global-scale yields are expected to increase or decrease (Parry et al., 2004; Nelson et al., 2009).

If CO_2 fertilization is strong, Europe and North America may benefit from climate change in the short term. Although, regions such as India and Africa are still projected to experience up to 5% losses by the end of 2050, even with strong CO_2 fertilization. These losses increase to up to 30% if the effects of CO_2 fertilization are omitted.

Without CO_2 fertilization, all regions are expected to experience a loss in productivity owing to climate change by the year 2050. Although, current global-scale studies have only utilized a limited sample of the available climate model projections (Parry et al., 2004; Nelson et al., 2009).

A reduction in CO_2 emissions would likely be reducing the positive effect of CO_2 fertilization on crop yields at a faster rate than it would moderate the negative impacts of climate change. Even if GHG concentrations rose no further, there is a commitment to a certain amount of further global warming (IPCC, 2007).

Stabilization of CO_2 concentrations would affect any increase in the impacts of CO_2 fertilization. On the other hand, the impacts of climate change could still continue growing. As a result, the impacts on global food production could be negative in the short term.

Nevertheless, predictions recommend that stabilizing CO_2 concentrations at 550 ppm would gradually reduce the production losses by the end of the century (Arnell et al., 2002; Tubiello and Fisher, 2006).

For all the species, the higher water-use efficiencies and major root densities under increased CO_2 in field systems may, in certain cases, increase drought pressures, still their large-scale implications are not well understood (Wullschleger et al., 2002; Norby et al., 2004; Centritto, 2005).

This could counterbalance some of the projected warming-induced increase in evaporative demand, as a result, easing the pressure for more irrigation water. In addition, this may also change the relationship between meteorological drought and agricultural/hydrological drought. An increase in meteorological drought may eventually result in a smaller increase in agricultural or hydrological drought owing to increased water-use efficiency of plants (Betts et al., 2007).

Soil moisture together with run-off may be much more relevant as compared to precipitation and meteorological drought indices as metrics of water resource availability, as these represent the water actually available for agricultural use. These quantities are regularly simulated by physically-based climate models as an important part of the hydrological cycle.

8.8.2. Ozone

It is evident that ozone is a major secondary air-pollutant, as ozone being at current concentrations has been shown to have major negative impacts, especially on the crop yields (Van Dingenen et al., 2009). On the other hand, in certain areas of North America and Europe, emissions of ozone precursors are gradually declining, while in other regions of the world, especially Asia, they are rapidly increasing (Van Dingenen et al., 2009).

Ozone, in general, reduces the agricultural yield via various mechanisms. First of all, it causes acute as well as visible injury to different products like in horticultural crops reduces market value. Secondly, ozone reduces the rate of photosynthesis and accelerates leaf senescence. This, in turn, impacts heavily on final yield of crops. Many studies have investigated such yield reductions (e.g., Morgan et al., 2003). Although, in various other regions, such as Asia, little evidence currently exists. Thus, the current understanding of the impacts in such regions is limited.

8.9. HOW TO PREDICT CLIMATE CHANGE IMPACTS?

Certain climate models have been developed that basically consider social as well as economic factors such as population and carbon emissions. By integrating these factors with a deep understanding of the regional as well as global climate science, various experts have developed such climatic scenarios that actually express the potential for diverse behaviors to impact climate patterns.

Climate scenarios have the capability of informing choices about the most probable impacts of precipitation, temperature, and seasonality on food production. Also, they allow people to guide different agricultural sectors on the best methods to adapt to different climate consequences by assessing the impacts along with the identification of tradeoffs.

The total effect of climate change on the global agriculture is not likely to be positive. However, there are few regions that will benefit, but most will not. As rise in atmospheric CO_2 concentrations are forecasted to stimulate the growth as well as improve the water use efficiency (WUE) in few species of the crops, climate impacts such as flooding, drought, and heatwaves mainly, will likely to reduce the yield potential.

While, indirect climatic impacts further include increased competition from weeds, expansion of pathogens and insect pest ranges and seasons, along with other alterations in crop agroecosystems. Adaptation strategies are short as well as long-term changes to the human activities, that eventually respond to the effects of changes occurring in climate.

In the field of agriculture, adaptation will need cost-effective investments in water infrastructure, emergency preparation for and response to extreme weather events, development of resilient crop varieties that deal with temperature and precipitation stresses, and new or improved land use and management practices.

8.9.1. Tools for Adaptation

Crop breeding for the development of new climate tolerant crop varieties is the main took for adapting agriculture to a changing climate. History along with the current experience of breeding shows that natural biodiversity within crops has allowed for adaptation of plants to different climatic conditions, providing clear evidence that plant breeding has literally a great potential to support the adaptation of crops to climate change.

Another tool is cropping system development, this can help agriculture adapt. For instance, the utilization of crop mixtures that have many crops growing at one time can aid systems exhibit high durability at the time of heat stress or high water.

- 8.9.2. What Are Our Research and Development (R&D) Needs?
- Development of drought-resistant crops that have been tested for potential yields when exposed to the periods of heavy water shortage.
- Improvements in plant nitrogen and water use efficiency along with development of cost-effective nitrogen uptake delivery systems and low-cost, high efficiency irrigation techniques.
- Development of global sites for testing and data collection as well as dissemination efforts, by making use of standard data protocols, to evaluate the performance of current as well as new genetic material and management systems in the current range of agroclimatic conditions.

- Continuous filed testing for tracking climate changes, breeding for resistance to new pests and diseases, and to deal with changes in pollinator distribution have been recognized as avenues to confront adaptation of crops in the name of climate change.
- Assessment tools development that include the biophysical constraints affecting agricultural productivity along with climate and socio-economic scenarios, including improved characterization of policy and program environments and options.

8.10. CONCLUSION

In the end, it is concluded that climate change is one of the most severe problems that are facing by farmers on an everyday basis. This is a prevalent problem that has been worsening as each year passes. With globalization and industrialization, there has been a significant increase in the carbon dioxide and greenhouse gases that directly contributes to climate change. The factors of climate change such as extremes weather, rising temperature, and a higher level of carbon dioxide considerably impact the overall productivity of the agriculture sector.

To make sure that the impact of climate change will be minimum, there is a need to engage in collective efforts by framing policies and strategies to help the farmer's communities. Some of them include looking for the factors that can directly impact the production, such as the use of pests and disease and change in water availability.

REFERENCES

1. Cho, R., (2021). *How Climate Change Will Alter Our Food.* [Online] State of the Planet. Available at: https://blogs.ei.columbia.edu/2018/07/25/climate-change-food-agriculture/ (accessed on 5 May 2021).
2. FutureLearn. (2021). *Impact of Climate Change on Agriculture.* [Online] Available at: https://www.futurelearn.com/info/courses/climate-smart-agriculture/0/steps/26565 (accessed on 5 May 2021).
3. Gornall, J., Betts, R., Burke, E., Clark, R., Camp, J., Willett, K., & Wiltshire, A., (2010). Implications of climate change for agricultural productivity in the early twenty-first century.*Philosophical Transactions of the Royal Society B: Biological Sciences, 365*(1554), 2973–2989.
4. Kim, D., (2007). *The Impact of Climate Change on the Agricultural Sector: Implications of the Agro-Industry for Low Carbon, Green Growth Strategy and Roadmap for the East Asian Region.* [Online] Unescap.org. Available at: https://www.unescap.org/sites/default/files/5.%20The-Impact-of-Climate-Change-on-the-Agricultural-Sector.pdf (accessed on 5 May 2021).
5. Mall, R., Gupta, A., & Sonkar, G., (2017). Effect of climate change on agricultural crops.*Current Developments in Biotechnology and Bioengineering*, 23–46.
6. Union of Concerned Scientists, (2019). *Climate Change and Agriculture.* [Online] Available at: https://ucsusa.org/resources/climate-change-and-agriculture (accessed on 5 May 2021).
7. White, K., Wagester, T., & Glasener, D., (n.d.).*How Will Climate Change Affect Agriculture*? [eBook] Available at: https://www.soils.org/files/science-policy/caucus/briefings/climate-change.pdf (accessed on 5 May 2021).

INDEX

E

F

G

H

I

K

L

M

N

O

P

R

S